MULLANE'S GUIDE
TO
LEARNING TO FLY

MULLANE'S GUIDE
TO
LEARNING TO FLY

Deep Dives into Select Topics for Today's Student Pilot

Patrick Mullane

Illustrated by Patrick Mullane
Cover Art by Daniel M.

ISBN: 979-8-218-33082-8

Also by Patrick Mullane:

The Father, Son, and Holy Shuttle:
Growing Up an Astronaut's Kid in the Glorious 80s

A Disclaimer:

While I hope this book provides helpful advice and interesting insights regarding learning to fly, it is not meant to be a flight instruction manual. You should consult FAA publications and seek instruction from a certified flight instructor if you wish to safely secure a private pilot certificate. Always consult FAA regulations, your pilot's operating handbook for the aircraft you fly, local airport procedures, and other official sources during and after your training. Adhering to regulations and decisions with respect to the safe execution of a given flight are ultimately and always the responsibility of the pilot in command.

Aircraft performance charts in this book are the creation of the author and meant to be illustrative; they are not to be used for real-world calculations.

Contact Me:

patrick@pjmullane.com

To my children

Table of Contents

Author's Note ... 12
Navigating This Book ... 15
Glossary of Acronyms ... 18
The Airplane's Parts ... 21
The Phonetic Alphabet ... 21
A Note on Names ... 22
Prologue ... 23
Chapter 1 I Hope I Don't Barf ... 27

- What do you do first? ... 29
- My discovery flight ... 32
- Why the left seat? ... 37
- It's what's on the inside that counts ... 39
- What's in a name? ... 42
- Turbulence: a problem for many ... 43
- Why you might be hooked ... 45
- Observation #1 – Flight uses all you've got ... 46
- Observation #2 – Flight celebrates human ingenuity ... 47
- Observation #3 – Flight connects you to others ... 48

Chapter 2 I'll Take the Beginner Package ... 50

- And now for some controversy … ... 52
- Do you have to be smart? ... 56
- Be ready to read … a *lot* ... 59
- How are a canal, flying, and Michelin stars related? ... 60
- Are you healthy? ... 64
- Always think happy thoughts ... 69

Chapter 3 How Safe Is This? ... 73

- But you're not flying commercial ... 74

But some good news … 78

Chapter 4 The Forces of Flight 81

The forces defined 83
Roll, Pitch, and Yaw 84
Drag and lift 85

Chapter 5 Why Planes Fly 87

So, what's the reason for lift? 93

Chapter 6 Why Planes Stop Flying - Stalls 94

Airspeed is only a proxy 98
The Spin Cycle 100
Why all this time on stalls? 105

Chapter 7 A Weighty Subject: Weight and Balance 108

V_a – The most mysterious airspeed 111
Balance 115
Calculating weight and balance 123

Chapter 8 Checklists: Cool Pilots Use Them 132

Walk-arounds and flows 135
The Invisible Gorilla 138
Don't be a deviant 141
Make the checklist work for *you* 145

Chapter 9 Can We *Go* Somewhere? 147

Scheduling and weather 147
Boredom 151
Measuring yourself 152
A quick note on scorekeeping 156
Teaching is hard 158

Chapter 10 The Written Test 163

Preparing for the written test 168

YouTuviators 171
Interlude: Fun after your PPL is in hand 174
My written test experience 174

Chapter 11 Some Important Aerodynamics 179

Slip sliding away 184
Side or forward? 190

Chapter 12 Solo Time 195

The IMPOSSIBLE TURN 202
Cleared for takeoff 205
The left turning tendencies 208
Shirttails and … spankings? 213

Chapter 13 Landings Part I: Pilots Manage Energy 215

Conservation of energy 215
Energy management when landing 218
Power in the pattern 222
Pitch and power, speed and altitude 224

Chapter 14 Landings, Part II: It's All About the Aim 230

Ballooning and sinking 234
The go-around 236
Fly all the way until touchdown … and parking 240

Chapter 15 Cross-Country Planning 241

Navigation 245
The sectional map 248
Course and Heading 252
Airspeed 258
Time to climb and time to descend 263

Chapter 16 Airspace 265

Class A You aren't going there (not yet) 266
Class B The BIG airports 266

Class C The middle child of controlled airports 271
Class D A great place to learn 273
Class E The most confusing of airspaces 274
Class G Fills in the gaps 281
Cloud clearance and visibility requirements 284
Special use airspace 293
MOAs – don't get shot down 293
Restricted airspace – the aliens may get you 295
Prohibited – the name means what it says 297

Chapter 17 The Solo Cross-Country Flight 300

The stage check 306
On my own 311
Good ADM … I think? 316

Chapter 18 The Weather and Atmosphere 322

Uneven heating at the local level 323
Stable air vs. unstable air 325
Pressure altitude – the path to density altitude 328
Density altitude – a short hop from pressure altitude 336
Why these topics? 341

Chapter 19 The Checkride 343

"Are you ready to go fly?" 346
The gouge … 347
Flight sims aren't always "games" 349
The government always wants its paperwork 350
The oral portion 352
Should we fly? 356
The flying part of the practical exam 358
V_x and V_y 362
Maneuvers time 367
Checkride landings 375

Epilogue 384

Acknowledgements 389
About the Author 392

Author's Note

All books are hard to write, even when you know with certainty the audience you are writing for. But they get doubly hard to write when the level of understanding of the topic being covered varies significantly among readers. That's the sort of audience this book is trying to address. You (I hope!) have an interest in aviation and becoming a pilot. But some of you are just beginning to investigate what it takes to become a private pilot; some of you have already soloed but have yet to pass your checkride; some of you are past the checkride and just want to be better versed in the important aspects of your new passion. Your own aviation experiences thus far will therefore drive your expectations for what you get out of these pages. Knowing this, I thought it was important to explain what this book is and what it is not.

First, this book is not an absolutely-everything-you-need-to-know-to-become-and-be-a-private-pilot book. Writing such a book would be an impossible task for any single individual. To illustrate the point, consider this: the FAA publishes three books you are encouraged to get your hands on when you begin studying to become a private pilot and those three tomes total over 2,100 pages. And there are things even those pages don't cover. While I'm not ambitious enough to think that I can out-do the FAA, I do hope that I have chosen topics that will interest you, inform you, and even entertain you.

If I'm not covering everything—or even attempting to do so—how did I decide what to include? I tried to focus on concepts that I found difficult to grasp while I was training or were so core to safe flying they had to be covered. I also focused on items I thought were most relevant "in the real world." After flying for two years with my certificate in hand, I felt better versed in what really mattered when it came to being a competent and safe pilot. I noted areas where I thought others would benefit from my learnings and focused there. I left out topics that are becoming more dated as technology advances and for which there is already much publicly available information to help pilots cement their knowledge. A good example of such a topic area is the use of VORs, a navigation aid. While you will be required to know how to use one to complete your training and get your pilot's certificate (as of the writing of this book anyway), the technology is becoming more dated in an age of GPS, particularly for your average visual flight rules (VFR) flyer. I therefore leave it to others to explain the navigation aid's proper use.

A second thing to keep in mind when it comes to the scope of this book is that the process by which you will become a pilot will be your own "snowflake" journey; no two people trace the exact same path to their pilot's certificate. Some of you will take longer than others. That might be because mastering the "stick-and-rudder" skills is hard for you, or it might be because you'll have a hard time understanding "book knowledge" you'll be tested on in a written exam. It may be both. Some of you are engineers. Some of you are musicians. Some of you are art historians. Some of you live in a place where the weather is perfect all the time (making frequent practice a breeze). And some of you live in Cleveland. There are tons of variables that will affect your journey relative to others. And where you are in your journey has a bearing on the number of concepts you've been introduced to and—more importantly—the number of them you grasp.

You may come upon sections in these pages that seem like review. If you do, take it as an opportunity to solidify learning you've already received from your flight instructor or through your ground school syllabus. As I've discovered in my own journey (and learning science confirms this), frequent review drives home learning. If you come upon a topic that is new, the same applies; power through it. There's nothing here that you won't have to know at some level to become a pilot, so you might as well start now.

To make the journey as entertaining and informative as possible, I do my best to include things you won't find in an FAA-produced book including interesting stories about the history and science of aviation and the story of my own journey through flight training. Flying is serious business a lot of the time but should be fun. So, if I do nothing else, my hope is that I introduce some levity every now and then that keeps you engaged while you learn. Fortunately for me, I have little competition in this area. I don't know if you're aware, but the FAA isn't known for its sense of humor. Enjoy.

Navigating This Book

Many of you may wish to jump to sections of the book that address a particular topic area. To that end, I've put the index at the beginning so that you can quickly access the information you need. For those just setting out to read the book, it will help you get a sense of what you'll learn in these pages.

ABCDE emergency acronym, 374
ADS-B, 146, 290
Adverse yaw, 103
Aft CG and rudder, 120
Age
  flight instructors, 33
  when starting lessons, 29
AGL
  difference from MSL, 149
Airman Certification Standards (ACS), 87
Airspace
  Class A, 266
  Class B, 266
  Class C, 271
  Class D, 273
  Class E, 274
  MOA, 293
  prohibited, 298
  restricted, 295
  special use, 293
Airspeed
  and stalls, 98
  calibrated, 261
  true, 261
  V_a, 111
  V_x, 362
  V_y, 362
Angle of Attack
  critical, 96
  defined, 94
ATC communications, simulator practice, 40
ATIS, 41
BC-GUMPS, 140
Books, for training, 51
Boredom, fighting it, 151
Center of gravity (CG), 117
Cessna vs. Piper, 32
Checklist origin, 133
Checkride
  author's checkride, 343
  gouge, 347
  maneuvers, 367
  oral portion, 352
  use of simulator to prepare, 349
Clarke's three laws, 23
Coordinated turns, 101
Cross country flight, choosing checkpoints, 250
Cross country flight, filling out log, 252
Cross country, author's flight, 311
Cross-controlling, 184
Datum plane, 118
Dead reckoning, 246
Discovery flight, 29
E6B, why it's unnecessary, 52
electronic flight bag (EFB), 54

Endorsements, use and importance of reading, 319
Energy
chemical potential, 217
kinetic, 216
management of, 215
potential, 216
Exam, importance of written, 164
FAR/AIM, 59
Female flight instructors, 35
First solo, 195
Flight following, 275
Flows, 136
Four forces of flight, 81
Go around, 236
Ground school, online, 168
Hotspot, 200
Humility, keeping you alive, 35
IACRA, 350
Impossible turn, 202
Inattentional blindness, 138
Instructor (CFI), picking one, 159
Invisible gorilla, 138
Isogonic line, 256
John F. Kennedy Jr. crash, 303
Journaling lessons, 155
KIAS, 108
Kollsman window, 262, 328
Landing
aim point, 230
energy, 215
flare, 232
roundout, 231
traffic pattern, 221
Leading edge, 88
Left seat for pilot, 37
Left turning tendencies, 209
Lessons, process for making more effective, 161
Levers and fulcrums, 116
Lift
and AOA, g-forces relationship, 112
and Newton's third law, 92
Bernoulli, 89
can't be explained, 91
Magnetic deviation, 256
Magnetic variation (a.k.a magnetic declination), 255
Maintenance, flight school, 39
Medical certificates, 64
Mental health, 69
Mental skills necessary to be a pilot, 57
MSL
difference from AGL, 149
N numbers, origin, 42
Nautical mile vs. statute mile, 259
Normalization of deviance, 142
PAPI, 220
Part 141 schools, 153
Part 61schools, 153
Pilot error and accidents, 78
Pilotage, 246
Pitch for airspeed, power for altitude, 224
Pitot tube, 180
POH - Pilot's Operating Handbook, 97
Preflight, 36
Private Pilot Certificate, 28
Private Pilot Knowledge Test, 163
Progess in training and measuring yourself, 152
Safety
flying vs. driving, 76
of GA flying, 73
Scorekeeping during training, 156
Sectional map, 248
Skids, 186
Slip
defined, 184
forward, 190

side, 190
Smart, do you have to be to be a pilot?, 56
Spins, 100
Stalls, 88
Stalls and fatal accidents, 105
Stick and Rudder, 92
TFR - temporary flight restriction, 251
Trailing edge, 88
True airspeed and air density, 262
True course, 253
True heading, 255
Turbulence, fear of, 43
Universal Coordinated Time (UTC), 294
V_a
and weight, 111
defined, 111
VASI, 220
VFR navigation log, 246
VFR rules, 148
V_x, 362
V_y, 362
Wake turbulence, 318
Weather, calculating pressure altitude, 333
Weather, cause of, 323
Weather, lapse rate, 326
Weather, pressure altitude, 328
Weather, stable and unstable air, 326
Weather, standard datum plane, 334
Weather, standard pressure, 334
Weather, temperature and density altitude, 336
Weather, thunderstorms and danger, 324
Weight and balance
calculation, 124
introduction, 108
weight x arm = moment, 123
Windshear, 180
Wing chord, 89
YouTube, as a training tool, 171

Glossary of Acronyms

ACS	Airman Certification Standards
ADM	Aeronautical Decision Making
ADS-B	Automatic Dependent Surveillance—Broadcast
AFB	Air Force Base
AGL	Above Ground Level
AIM	Aeronautical Information Manual
AIRMET	Airman's Meteorological Information
AOA	Angle of Attack
AOPA	Aircraft Owners and Pilots Association
ASI	Aviation Safety Inspector
ATC	Air Traffic Control
ATIS	Automatic Terminal Information Service
CDI	Course Deviation Indicator
CFI	Certified Flight Instructor
CFII	Certified Flight Instructor Instrument
CFIT	Controlled Flight Into Terrain
CFR	Code of Federal Regulations
CG	Center of Gravity
Critical AOA	Critical Angle of Attack
CTAF	Common Traffic Advisory Frequency
DPE	Designated Pilot Examiner
E6B	Designation for Slide Rule Flight Computer
ECAC	East Coast Aero Club

EFB	Electronic Flight Bag
FAA	Federal Aviation Administration
FAR	Federal Aviation Regulations
FPM	Feet per Minute
FSDO	Flight Standards District Office
FTN	FAA Tracking Number
GA	General Aviation
HS	Hotspot
IACRA	Integrated Airman Certification and Rating Application
ICAO	International Civil Aviation Organization
IFR	Instrument Flight Rules
IMC	Instrument Meteorological Conditions
KCAS	Knots Calibrated Airspeed
KIAS	Knots Indicated Airspeed
MEI	Multi-engine Instructor
METAR	Meteorological Aerodrome Report
MOA	Military Operations Area
MSL	Mean Sea Level
NTSB	National Transportation Safety Board
PAPI	Precision Approach Path Indicator
PHAK	Pilot's Handbook of Aeronautical Knowledge
PIC	Pilot in Command
PIREP	Pilot Report
POH	Pilot's Operating Handbook
PPL	Private Pilot's License
RPM	Revolutions per Minute
SIGMET	Significant Meteorological Information
SUA	Special Use Airspace

TAF	Terminal Area Forecast
TAS	True Airspeed
TFR	Temporary Flight Restriction
TFR	Temporary Flight Restriction
VASI	Visual Approach Slope Indicator
VFR	Visual Flight Rules
VOR	VHF Omnidirectional Range

The Airplane's Parts

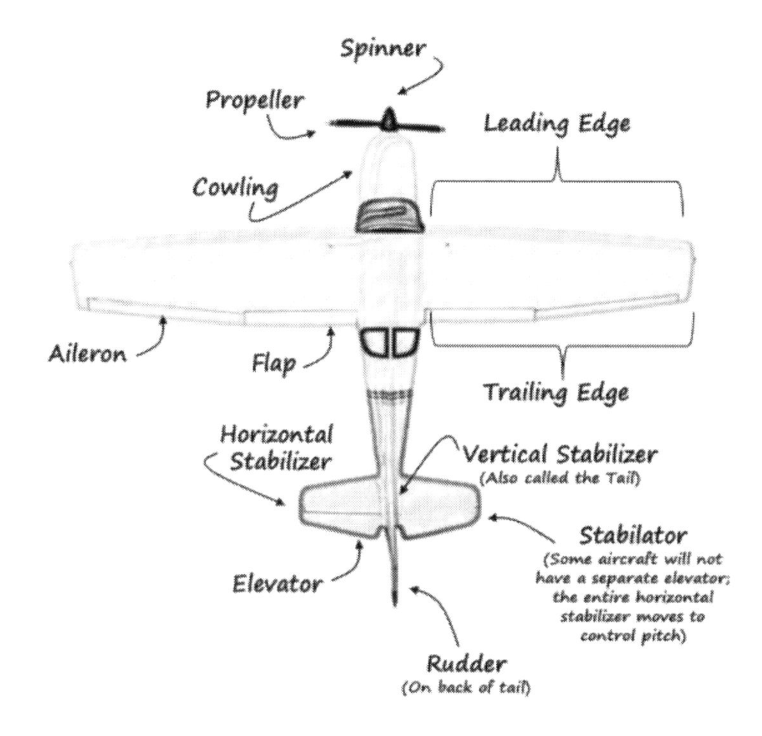

The Phonetic Alphabet

A=Alpha	I=India	Q=Quebec	Y=Yankee
B=Bravo	J=Juliet	R=Romeo	Z=Zulu
C=Charlie	K=Kilo	S=Sierra	
D=Delta	L=Lima	T=Tango	
E=Echo	M=Mike	U=Uniform	
F=Foxtrot	N=November	V=Victor	
G=Golf	O=Oscar	W=Whiskey	
H=Hotel	P=Papa	X=Xray	

A Note on Names

The names of my flight instructors in this book and the Designated Pilot Examiner (DPE) who gave me my checkride have been changed. There is one exception. Jim Henry, CFII and MEI, facilitated my stage check ride in preparation for my checkride. He is referenced in Chapter 19.

Prologue

"The aeroplane will never fly." So said Lord Haldane, Britain's Secretary of State for War. Looking at a picture of Haldane today, it's hard not to believe everything he says. He looks distinctly British—stiff upper lip and all that—and you imagine him saying the word *aeroplane* in a baritone voice laced in phlegm. But here's the interesting thing: not only was Haldane wrong, but he made his pronouncement in 1907, four years *after* the Wright brothers flew at Kitty Hawk.

Now before you get too down on Lord Haldane, it's important to note that he was not alone in his incredulousness. Skepticism with respect to successful heavier-than-air flight was common in his time and the secrecy of the Wrights didn't help. In 1906, the *Paris Herald* noted with an almost ambivalent tone, "The Wrights have flown or they have not flown. They possess a machine or they do not possess one. They are in fact either fliers or liars. It is difficult to fly.... It is easy to say 'We have flown.'" Apparently, concern about "fake news" preceded the internet and social media.

More than a century since Haldane's statement, flight is still the sort of thing that seems impossible to be real. In my first book, *The Father, Son, and Holy Shuttle: Growing Up an Astronaut's Kid in the Glorious 80s*, I referred to Clarke's Three Laws, and I find it appropriate to cite them again here (thus demonstrating that the laws are quite versatile, at least to me). The "Clarke" referred to here is Arthur C. Clarke, one of the world's most gifted and famous science fiction writers. In his books and essays about the future, Clarke made three assertions that were eventually codified into what are now known as his three laws. They are:

1. When a distinguished but elderly scientist states that something is possible, he is almost certainly right. When he states that something is impossible, he is very probably wrong.
2. The only way of discovering the limits of the possible is to venture a little way past them into the impossible.
3. Any sufficiently advanced technology is indistinguishable from magic.

Lord Haldane's inaccurate statement with respect to flight bumps up squarely against laws number one and three (although Haldane was not a scientist). Haldane was wrong that something was impossible, and he was probably wrong because the idea of humans buzzing about the air in controlled flight seemed like magic. So magical did it seem, some still thought it fake news well after reports of the Wrights' success began go show up in newspapers.

I, for one, *still* think it seems like magic. I admit, though, that it's sometimes hard to maintain that wonder. Most of us experience flight as a litany of long lines, kicks to the back of the seat, cramped spaces, and ridiculous ticket rules. But step back a minute. *Really* step back and get some perspective. Consider this: in 1867, a trip from New York to San Francisco on an express train took 83 hours. And that was considered blazingly fast. One hundred years later, you could do it in about six hours on an airliner.

Speed isn't the only miraculous thing about human flight. It may be hard to believe this, but you make your transcontinental flight in incredible comfort, even if the guy next to you is spilling into your seat like beer foam breaching the rim of a glass. You have a pressurized, temperature-controlled cabin and are served nuts and soft drinks, or, if you prefer, one of those beers that breaches the rim of a glass. You can check your email. You can watch live television. You can even belly up to a bar. That's right, a bar. I did this once. I had the good fortune of flying on an Airbus A380 (that double decker monster of the air) from Mumbai to Dubai. The back of the plane on the second level had a u-shaped bar like one you might find in the lobby of a boutique hotel. There I was, an American, on a European-made airplane where I ordered, from a South African bartender, an alcoholic beverage from Scotland all while I gazed at a map showing an icon of our aircraft over the Gulf of Oman passing off the coasts of Pakistan and Iran at 38,000 feet. I covered the roughly 1,100 nautical miles in three hours. How, in God's name, is that possible?

Yes, flight is still magical. And those of us who want to be pilots-in-command of aircraft rather than just passengers on them, know this. It is the single common thread that runs through all of us. It is in us that the majesty and wonder of one of the most amazing technological feats in human history lives; indeed, we keep it alive.

This book is a few things. First, it is my own story of chasing a lifelong dream to become a private pilot. Second, it is an eclectic sampling of the history and science of aviation that pays homage to those who came before us. Finally, and most importantly, it is instructional in nature. As I already mentioned in the author's note, it is not an exhaustive how-to book. But I hope I offer a wealth of useful information for those interested in, or already pursuing, a private pilot's certificate. That said, I'm not a CFI (Certified Flight Instructor); if you want to learn to fly, you need to find one of those. But for those who have yet to contact a CFI, I hope this prompts you to do so. For those who already have found their instructor and are knee-deep in training, tired of hearing "more right rudder!" yelled at them, I hope these pages keep you motivated and perhaps give you nuggets of knowledge that can serve as conversation starters with your CFI. Finally, for those already commanding the air with a private pilot's certificate in their back pocket, I hope the book is an entertaining and informative walk down memory lane as well as an important dive into several key concepts—concepts that, once better understood, can help you be a better pilot. I guarantee you'll learn new things. After all, how many of you knew of Lord Haldane?

There is probably no community that thrives on storytelling more than the aviation community. When among our kind, man or woman, young or old, we become the airborne descendants of hunters in a circle around a fire reliving the kill of the day, our flat hands modeling an aircraft in flight: *There I was* .... This is my contribution to that circle of trust, a circle of people from all walks of life who share a connection based solely on the thing Lord Haldane told us we would never be able to do. It's time to fly....

Chapter 1
I Hope I Don't Barf

For many destined to fly, the seed to pursue aviation is planted early. Maybe your quest to become a pilot began with a visit to the cockpit, a visit like that of Joey, the young boy who toured the flight deck in the movie *Airplane!* In the famous scene, Joey stands between Peter Graves (Capt. Clarence Oveur) and Kareem Abdul-Jabbar (copilot Roger Murdoch), eyes wide with wonder at the machine and those that fly it. But if you did get to visit a cockpit as a child, I'm hoping that you got to ask more questions than Joey. It shouldn't be hard. That's because Joey didn't get a chance to ask a single one during his visit. It's Captain Oveur who does the questioning. And, if you've seen the movie, you know those questions are … unusual.

If your interest in aviation was stoked by *Airplane!*, I won't judge. But I think it's a bit more common to cite *Top Gun* (loved it), *Sully* (not my favorite), or *Flyboys* (cool World War I scenes even if within a hokey plot).

Or, maybe, it was a family connection.

That's what it was for me: a family thing. My father was an Air Force aviator who became an astronaut when the first group of space shuttle astronauts was selected in 1978. While having a dad who flew in space was undoubtedly cool, I bought into the dream of flying long before that. When we were just three or four years old, Dad would take my twin sister and me out to the perimeter road around a runway at the base where he was stationed in England, and we'd stand beyond the threshold about 150 feet from F-4 Phantoms in full afterburner before they released their brakes and thundered into the distance. I swear that noise, the heat, and the kerosene in the jet fuel seeped into my bloodstream through my skin and infected me with a lifetime of aviation illness. From those early days I drew airplanes, read about airplanes, and built model airplanes.

While interest in aviation early in life seems to be common in the community of those who want to learn to fly, it's by no means universal. I have a friend who had never really considered becoming a private pilot until his own son expressed a desire to fly. My buddy reluctantly joined his son on a discovery flight (more about those later), and it seemed to have the same effect on him that the runup of an F-4 did for me so many years before. He was hooked.

On an internet message board where somebody asked others where their interest in flying came from, many "late bloomer" stories abound. One guy didn't get his private pilot license (PPL)[1] until well into his professional career after growing frustrated by the hassle of commercial travel. Becoming a private pilot for him was—at least initially—very utilitarian. But in his post, there's a hint of a tone that indicates that he, like most of us, went "all in"

---

[1] "Private Pilot License" is technically not the right term for the credential you earn to legally fly an aircraft with somebody else on board. Rather, it's called a "Private Pilot Certificate." But PPL is so commonly used in the aviation community, I will use it in these pages.

after he got a taste of flying. Another responder to the question said that they had no interest in becoming a pilot until, as an adult, they watched the documentary series *Air Crash Investigations* on the National Geographic Channel. Rather than be turned off by the bad things that could befall an aircraft, this "glass half full" person was fascinated by the discipline and structure it took to command an airplane and turned that fascination into a passion, a passion that likely exceeded many with an earlier start.

I note these examples because if you did have a later epiphany regarding learning to fly, you could find the intensity of those who were all-in when barely out of diapers a bit intimidating. It can lead you to feel "behind." But I've learned there's little correlation between when you unearth an interest in flying and your ability to complete training and become a proficient pilot. And, in fact, there are plenty of early bloomers who, once they begin their journey, realize that it wasn't what they had imagined and walk away from their dream like somebody walking away from a bad first date. So, if you're over the age of forty and are just starting to think about getting a private pilot certificate, don't beat yourself up wondering if it's too late. It's not. You'll be just fine.

What do you do first?

If you've never experienced flight in a single-engine, piston aircraft (more commonly known as a "light" airplane), the first step you should take is to book a discovery flight at a local flight school. The name of the flight implies what it's for—to help you discover whether this flying thing is for you. Just use Google; typing "discovery flight" into my search bar a moment ago yielded three options within an hour's drive of my house. Admittedly, I live in a large metropolitan area with many airports in the vicinity. In more rural areas, the number of options may be significantly more limited.

Note that since it's likely you will do your discovery flight at a flight school, you should also have a secondary objective beyond the flight itself: learning about the school. After all, if the discovery flight piques your interest, you will want to continue with lessons, and the place where you first get your hands on the controls would be a logical place to continue to work your way toward your PPL.

I started thinking about a discovery flight in the winter of 2021. My youngest child had graduated from college that same year and so I suddenly felt like I had gotten a big pay raise. Flying is not a cheap hobby, and having those dollars previously earmarked for tuition suddenly available for other uses revived a long-held itch to climb into a cockpit. When I did my own Google search, I found a large school at Hanscom Field, a busy airport about 13 miles northwest of Boston and about 30 minutes from my home. Hanscom also is the name of an Air Force base co-located with the civilian operations.

It's a little-known fact that a large majority of Air Force bases are named after people killed in plane crashes. This tradition seems a bit macabre to me—like naming hospitals after patients who die in them. But the tradition is deep. Edwards Air Force Base (AFB), the U.S. Air Force's premier flight test base, is named after Captain Glen Edwards, who was killed while flying the YB-49 bomber, a jet-powered "flying wing" that, despite my love of aviation, if you'd ask me to get into it my response might have been something like, "Hell no." It *looked* like it shouldn't fly. Andrews AFB, best known as the home base of Air Force One, is named for Lieutenant General Frank Maxwell Andrews who was killed in the crash of a B-24 Liberator bomber during World

War II. MacDill AFB in Florida is named after Colonel Leslie MacDill, an aerial gunnery school commander killed when the airplane he was in had an engine failure in 1938.[2]

Hanscom is no exception to this naming convention but is a bit different in one regard. While Laurence Hanscom was indeed killed in a plane crash, he was a civilian, a reporter, and aviation advocate who had lobbied for an airfield to be established in Bedford, Massachusetts. It's a good thing for me he did that advocating in the 1940s because there is absolutely no way the airport would have been built today. While Bedford is not a memorable town name to most Americans, it sits squarely between two hamlets much more familiar to anybody who's studied even a little American history: Lexington and Concord, the birthplaces of the American Revolution. In fact, paralleling Runway 29/11 at Hanscom and just a little over a mile to the south is a national park that snakes from the east to the west, marking the path the British regulars took from Boston to Concord through the town of Lexington (it's also the route the redcoats withdrew on after the "shot heard 'round the world" was discharged in Concord, taking heavy casualties along the

---

[2] Unlike military bases, civilian airports in the U.S. are almost always named after politicians. Edward Lawrence Logan was a military leader himself but was better known in the Boston area as a politician. A quick scan of other major airports in the U.S. shows the use of politicians' names is rampant: Atlanta Hartsfield (mayor), New York Kennedy (president), Houston Hobby (governor) and Houston George Bush Intercontinental (president), New York La Guardia (mayor), Washington Dulles (secretary of state), Las Vegas McCarran (senator) renamed in 2021 to Las Vegas Reid (senator) … and on and on. It's an irony that one exception to this tradition is in a town known for its political machines, Chicago. Its main airport, O'Hare, is named for a Medal of Honor recipient.

way). Just outside the main road into the flight school where I did my discovery flight is a stone marker in the national park denoting the location that Paul Revere was captured by the British while on his famous ride in 1775.

My discovery flight

It was through Minute Man National Park that I drove to get to East Coast Aero Club (ECAC) for my discovery flight one day in late February of 2021. COVID had the country constipated—not much was happening anywhere—but flight instruction had been deemed essential by the federal government. It was beyond me why this was the case, especially since air travel had become significantly curtailed during the pandemic and flight training required a flight instructor and student to sit shoulder to shoulder in a cramped, enclosed space. About the only way to introduce more risk of transmission was to ask two people to kiss—and no offense to my instructor, but I wasn't interested. That said, this may be a case where the Feds had more foresight than I did. As air travel came back through the summer of 2021, the shortage of pilots the airlines experienced was critical. Keeping the pilot pipeline going during the height of the pandemic helped relieve this shortage and while it did so only modestly, it would have been much worse if flight instruction hadn't been allowed to continue.

I had scheduled my discovery flight on a previous visit to ECAC after taking some time to find its offices in an old, 1950s era civil air terminal. Like many government buildings of the period, it had a bit of a prison vibe. Utilitarian to the point of austerity. By contrast the staff at ECAC was bright and cheery and set me up for a ride in a Cessna 172. If you're reading this book, you probably already know of the 172, the workhorse of global flight training. But ECAC's fleet was comprised mostly of Piper Warrior aircraft, not the ubiquitous 172. It was nice to have options to choose from. But you may not. Don't worry though, either aircraft (and a few other models as well) are great trainers. The primary difference between the Piper and Cessna aircraft is the location of the wing; it's above the fuselage on the Cessna and below it on the Warrior. Like all things related to engineering and design, these different configurations offer tradeoffs. But the primary difference is one of visibility. If you like to be able to see easily above and beside you, the Piper design is better. If you'd like to be able to look straight down at the ground passing under you, the Cessna is your choice.[3]

Why did I choose to take my discovery flight in a Cessna 172? The primary reason was that I had a dream of one day buying an airplane, and the wide availability of the 172s was appealing. They are well known to mechanics. Parts are easily available. They are stable and safe. If you don't buy and want to rent when traveling, the odds are that you'll find 172s available for hire. All these things pointed me to doing my training in a 172, and so I thought that I might as well make my first flight in the aircraft I planned to finish my training in. Later, though, the broader availability of the Warriors at ECAC led me to transition to them for the rest of my training.

The roster of instructors at my school was exceptionally large by flight school standards—somewhere between 20 and 25. I

---

[3] Cessna 172s are also easier to get in and out of, with doors on both sides of the aircraft. While this wasn't a big deal to me during my training, after I got my certificate and began flying with my wife, she really didn't like the single door on the Warriors.

didn't know anybody to ask for, so an instructor was randomly assigned to me. Jeff was a former military helicopter pilot who had been in the business world for a while but decided to pivot to an airline job. He was using flight instruction to "build time"—i.e., gaining flight experience while earning money so that he had enough flight hours to be eligible for an airline position. This is the norm among flight instructors and makes the student-pilot relationship one of the most bizarre in all of teaching and learning. This is because instructors tend to be young. Jeff was probably around 30 … and he was *old* relative to other instructors at the school. When Jeff was later hired to the airlines during my training, my second instructor was even younger. By contrast, almost all the students were, like me, middle-aged. The cost of flight lessons means many students must have some level of professional success (or a rich relative) to be able to afford training.[4] I know of no other industry where the young teach the "old" on such a large scale. This can create some weirdness; young instructors may be afraid to be critical of older, successful students and some students can be know-it-alls, thinking their professional success makes them experts in everything.

To that last point (and a bit of a lecture): don't be that "overconfident" student, otherwise known as an a**hole. Just because you have had success in virtually all other areas of your life doesn't mean you will be a good pilot. It's not only rude and

---

[4] For those of any age who don't have the resources to pay, be sure to really investigate all options. Some organizations offer scholarships to aspiring pilots, many airlines have begun flight academies to train their next generation of pilots, and some flight schools give opportunities to those who work the ramp for them. It's important to take control of your own dream and "beat the bushes" to find a path that will work for you.

disrespectful to behave with arrogance toward your instructor who may already feel awkward about telling you what to do, but it's dangerous. Humility literally keeps pilots alive. I'd argue it's the *single* most important aspect of safety. Learn to be humble in the cockpit early on, and you'll significantly increase the chances that you won't be killed doing something stupid in an airplane.

After Jeff and I exchanged pleasantries, he asked if I'd mind if Brittany, a new instructor at the school, joined us on our flight. It was to be her first flight as an employee of the school, and she was learning how they did things. Brittany is a rarity in the aviation world. According to Women in Aviation International, just 7.3% of flight instructors are women. This is too bad. We need more women in aviation. The good news is that the percentage of female pilots is increasing, if slowly. If you are a woman reading this book, I'd encourage you to find a network of other women who enjoy aviation as you do. One place to start is with The Ninety Nines (its name is a reference to its 99 charter members). It's a nonprofit group of women pilots with a mission to advance aviation among other women. Their first president was Amelia Earhart, so it's hard to argue with their bona-fides! The Ninety Nines has chapters all over the world. Find one if you haven't already. They, like Women in Aviation International, are among the organizations offering aviation scholarships to their members.

Jeff, Brittany, and I walked across the tarmac out to the plane. I loved that walk across the flight line. Still do. There's something about being "inside the perimeter." It's silly, I know. But I have a childlike wonder at walking among flying machines. It's an even more pronounced wonder at Hanscom where $50 million airplanes are coming and going alongside our little single-engine piston "toys." It's easy to see such sophisticated equipment and wonder how anybody masters all they must to fly them safely. But remember every pilot who commands those aircraft started as you did—in a single-engine trainer of some kind. It's easy to forget that when you become overwhelmed by all you'll need to know and do during your initial training.

We reached the airplane and began a "preflight," an inspection of the aircraft to ensure it's ready to safely take to the air. Later I learned the axiom to "do your preflight as if your life depends on it … because it does." And indeed, that's true. The history of aviation is littered with stories of pilots getting hurt or killed because they didn't do the obvious: ensure there was fuel in the tanks, remove protective covers from critical external instruments, check the oil. Unfortunately, like a lot of things in aviation, it's easy for the exercise to become rote; you do it, but your mind is on autopilot and not really registering what your body is doing. Did you really take note of the oil level? Or just remove the dipstick to check a box? Complacency of this sort is the hardest thing to keep from creeping into all that you do as a pilot. Every now and then, remember the feeling of your first preflight, when you counted every rivet, and remind yourself to recapture it later in your career when you catch yourself just going through the motions.

With the inspection of the aircraft complete, Brittany crawled into the back seat and Jeff and I took our places in the front. I slid into the left seat, considered the "pilot in command" seat, and Jeff took the right. This is common in a training environment; you will take the seat you will fly from when you have your PPL even though the right seat in all trainers has a set of controls too. You have probably noticed this is a custom at all levels of aviation. The captain on an airliner also sits in the left seat, the copilot in the right. The commander ot of a space shuttle sat in the left seat, the copilot the right.[5]

Why the left seat?

But *why* does the pilot sit in the left seat instead of the right if the controls are the same and the instruments are in view? While researching this I found many theories, and none seem to percolate from the stew of conjecture to the certainty of fact. It seems nobody really knows for sure. The Wright Brothers' first airplane with two seats had the controls on the left. So, some say it all started with them. Of course, that begs the question of why *they* put the controls there.[6] I could find no information to answer that question. Some think the left-seat convention comes from a desire to organize the cockpit for the 90 percent of us who are right-handed. Since all the work inside the cockpit not related to manipulating the stick or yoke (the control a pilot uses to maneuver the airplane) is instead related to configuring radios and other avionics or manipulating the throttle and mixture, the theory is that all those fine-motor skill activities are best put to

---

[5] While the person in the left and right seats of a shuttle effectively do the same tasks as a pilot and copilot, respectively, it's interesting to note that NASA called the left seat occupant "Commander" and the right seat occupant "Pilot." This confusing convention is apparently born of ego. No astronaut wants to be known as "copilot."

[6] It is logical to assume the Wrights put their controls on the same side that automobiles in the United States had their steering wheels. But the timing doesn't work. A July 2021 article in *Plane and Pilot* magazine points out that the Wrights' first left-side controlled airplane flew *before* Ford standardized the left-side steering wheel in the Model T in 1908. And, in fact, most cars before the Model T had a steering wheel on the right.

the right of the pilot by putting him or her in the left seat. I'm skeptical of both explanations.

The theory I'm most bought-in to is one related to the physics of flight, which all aspiring pilots will learn about early in their training. Engines in single-engine, piston-driven aircraft spin the propeller clockwise when viewed from the cockpit. For several reasons we'll get into in more detail later, this single fact leads the aircraft to want to naturally turn to the left. This left-turning tendency is surely something early aviators knew well as soon as a single propeller was introduced to the design of early airplanes. (The Wrights didn't have this issue as their craft had two propellers spinning in opposite directions effectively muting the forces a single-engine aircraft experiences.)

This is all very interesting, I'm sure you're thinking to yourself, *but why would a left-turning tendency drive the placement of the pilot's seat on the left side of the airplane?* The theory goes that during that most critical phase of flight, landing, when flying in the traffic pattern around an airport (for the uninitiated, the "pattern" is a rectangular track around the runway that pilots fly to set up for a landing), you might as well turn in the direction the airplane already wants to turn. Indeed, a "standard" traffic pattern is made up of left turns; the FAA even chimes in on this and says that unless otherwise directed, you should fly a pattern with left turns only. And if you are going to be turning left, you want to be on the left side of the plane as it gives you a better view of the runway, allowing you to set up for your approach more precisely.

I don't know the degree to which this really is the reason for a pilot-in-the-left-seat convention. But I do know this: it does indeed feel more comfortable to make traffic pattern turns to the left. At the busy airport where I've learned to fly, I often must make right traffic at the request of the tower, and it does feel a bit like rubbing your stomach while patting your head. I also find myself craning my neck to look over the instructor or passenger in the right seat to get a view of the runway, all the while ensuring I'm properly counteracting the tendency of the plane to turn to the left, away from the runway.

It's what's on the inside that counts

When I did take my place in the Cessna's left seat, one thing that surprised me was just how old it looked. Paint was worn, metal was scuffed, seats were creaky, and fabric was torn. I couldn't help but think that Charles Lindbergh might have sat in the very seat I was sitting in. Be prepared for this. Trainers take a beating and, in most cases, have been doing so for decades before you sit in them. In fact, N488BA (the call sign of my discovery flight aircraft) was first certified for flight in 1979, when I was 11 years old and *Top Gun* was still seven years from release. In fairness to ECAC, this was one of their older planes and they aren't unique in having some dinosaurs in their fleet. But even in the case of younger models, it's very rare for an airplane at a busy flight school to look brand-new. Fortunately, cosmetics aren't very important to airworthiness. An airplane will fly with chipped paint. What's more important is the upkeep of the engine.

This brings me to something you should ask your flight school about: How do they maintain their airplanes? Do they have mechanics on staff? Do they contract out their maintenance? Do they have documentation you can review that verifies they are doing the required inspection regimes (they better!)? The FAA requires strict upkeep of aircraft maintenance records; you shouldn't be afraid to ask to see them. At ECAC, there is an electronic board in the main office showing all the aircraft and the status of inspections and maintenance efforts. Things like this can help give you some level of comfort you are flying a well-maintained plane. Remember: a torn side pocket won't kill you; a cracked crankshaft can.

With the engine started and the radios turned on, our headsets came alive. I asked Jeff if he'd let me do the radio calls to ground controllers to get permission to taxi. This was, I'm sure, an unusual request for a discovery flight—and Jeff seemed a bit unsure—but I had been practicing air traffic control (ATC) communications using a home flight simulator and an internet-based service that allowed me to tap into a virtual world of airports and talk to real humans who served as controllers, acting and speaking just as controllers do in the real world. This practice was invaluable. Tutorials on the service's site provided training for a given mission and badges were earned if you were able to communicate correctly during that mission. For less than one-fifth the cost of an hour of airplane rental, you get a month of virtual practice. If you don't do something like this, you will have to do all your learning while in the cockpit, with the engine running, fuel burning, and the instructor billing you for the time. And while a controller might yell at you in the virtual world (this happened to me), you won't be reported to the FAA if you do happen to screw something up.

Jeff asked me what I was going to say to the ground controller, understandably skeptical that I was prepared to take on communications. I told him: "November 488 Bravo Alpha, at the west ramp, ready to taxi with Charlie."[7] "Charlie" referred to

---

[7] In aviation, the phonetic alphabet is used to designate letters. Rather than saying the sound of the letter *A*, you say Alpha. Rather than *B*, you say "Bravo," and so on. This is to prevent confusion over crackly radios, where

the letter *C*, which in turn denoted a specific ATIS report. ATIS stands for "Automated Terminal Information Service." The service consists of a recording created by somebody in the tower each hour that provides standard information to pilots departing and arriving the airport. It includes information such as sky conditions, runways in use, altimeter setting and much, much more. Each hourly report is given a letter to ensure that pilots know which one is current. There's not much sophistication to this. The next hourly report after Charlie would be Delta. Pilots can listen to the recording on an assigned frequency, helping ensure that radio communications on the tower frequency are not cluttered and repetitive in nature.

Jeff seemed satisfied with my radio communications rehearsal, and so he let me make the call. Even with all the practice I'd had in the virtual world, and the short practice I'd just had telling him what I would be saying to the controller, there was a hesitation before I depressed the mic button on the yoke. I've learned this is very common among new pilots. For some reason, pushing a button to speak makes your brain turn to mush, at least early in your training. I still screwed up the call, although I can't for the life of me remember how. I've since learned that most controllers are kind and patient, especially at an airport where there is a lot of training going on. And on that day, the controller I spoke to responded as if I had nailed the call. I know it probably won't matter that I say this to those of you who are just starting out, but don't get nervous about the radio. Talking over the airwaves is just a conversation made up of very short, prescribed exchanges. Admittedly, those exchanges are in their

---

A may sound like "eight" or *B* may sound like "three." The full phonetic alphabet is provided at the beginning of the book.

own sort of language. But when all else fails, just speak in plain English. I've done this several times during my training and after getting my PPL. In each case, the controller was helpful.

What's in a name?

One thing you may have wondered about is why there a *N* at the beginning of all aircraft identification numbers (as in the aircraft I flew on my discovery flight: "N488BA")? Well, *all* aircraft don't have an *N* at the beginning, just aircraft registered in the United States. In Great Britain, these identifiers begin with a *G*. In Germany they are *D* (as in Deutschland). In Japan, *J*. Wait a minute … why do these countries get obvious letters to begin their aircraft identifiers and the United States, where the airplane was invented, gets *N*? Why not *U* or *A*? According to some history available on the FAA website, you can blame that on the United States Navy (*N*=navy) which used the designation for various radio-related operations as early as 1909. Therefore, when, in 1919, the International Air Navigation Convention doled out letter designations for aircraft identification, the U.S. was given *N* (since it was already in use by the U.S. Navy), *W*, and *K* (which were apparently completely random assignments). The United States government reserved the use of *N* for itself and began assigning this prefix to aircraft registered in the country. It then decided to give the *W* and *K* designators to radio stations for use as their identifiers. That's why, to this day, the call letters of radio stations in the United States begin with *W* if east of the Mississippi River and *K* if to the west.

While the history of U.S. aircraft identifiers is fascinating, it wasn't top of mind for me as Jeff, Brittany, and I were cleared to the active runway, taxied there, and did some final checks. Jeff let me handle the aircraft as much as possible, even letting me pull back on the yoke to get the plane into the air. After that, we headed southwest of the field to a practice area and did basic maneuvers, again with me manipulating the controls. While there was a bit of wind that was noticeable on takeoff, at about 2,500 feet it was relatively smooth. A blanket of snow covered the land as far as the eye could see and the sky was, as you'll often hear aviators say, "clear and a million," the "million" referring to miles of visibility. We were fortunate to have a smooth flight. For most parts of the United States, such smoothness is, if not rare, then certainly a gift when you get it. The troposphere, the "bottom" layer of atmosphere that we live and fly in, is a tumultuous place. We'll talk in more detail about why that's so in a later chapter on weather and the atmosphere. But suffice it to say that turbulence has a bumpy relationship with many flyers.

Turbulence: a problem for many

Among those embarking on their PPL journey, fear and nausea caused by turbulence are common. Turbulence by itself, except in extreme cases, is a non-issue for all types of aircraft. Yet it's by far one of the most common anxieties identified not just by prospective pilots but by those who fly as passengers on commercial airliners. My own wife has this anxiety. Like many, she worries the wings will come off the airplane. I think this is because most envision turbulence acting *only* on the wings. The cabin seems fixed relative to our own position, and so our brains tell us the wings are taking the brunt of the turbulence while the cabin stays stationary, as if the wings were airborne shock absorbers riding a rocky road of air. This sensation is reinforced by a look out the window of an airliner where you can often see the wings flexing in the middle of a turbulent ride. But the entire plane is rockin' and rollin'. Afterall, if that wasn't true, then you wouldn't feel anything in the cabin when the plane hit turbulence. And don't worry about those flexing wings. Modern airliners can withstand incredible stresses. A *Wired.com* article from 2018 discussed Boeing's wing flex test on a 787 Dreamliner. In a picture accompanying the article, the wings of a test aircraft in a hangar have been bent upward 25 feet beyond horizontal, well beyond the most extreme forces an aircraft will ever be subjected to in flight.

Like a lot of things in your early training (and a lot of things in life for that matter), overcoming discomfort comes with repeated exposure to discomfort. This is true with respect to turbulence's evil twin, nausea. If you experience nausea on your discovery flight or during the several flights after that, don't give up. My father, whom I never remembered getting sick no matter how much a plane was bouncing or a boat was rocking, told me that early in his Air Force career he puked his guts out on every flight. And he went on to become an astronaut. Eventually, the body adjusts. Part of the adjustment is certainly physiological, but there's also a mental aspect to it. As you get deeper into your training, your mind will be occupied, and it will be you who will be telling the aircraft what to do. Keeping the brain from thinking about your stomach and having some knowledge of the forces you'll next feel on your body—because you're the one commanding the aircraft—goes a long way toward overcoming any early nausea.

After about 45 minutes of doing turns, changing altitude, getting an overview of the instruments, and other "beginner" stuff, we headed back to the field. I had my hands on the controls until we got to the traffic pattern and then Jeff took over to bring us into our landing. At an airport like Hanscom where business jets are coming and going with high regularity, final approach is another area where smooth air is an incredible rarity. Larger, heavier aircraft leave a wake of roiled air behind them.[8] But, again, don't be surprised by some bouncing around if you are flying at a similarly busy airfield with traffic of varying sizes.

Why you might be hooked

I had gone to Hanscom on that day not intending to immediately begin flight training. While funds had been freed up by my kids getting out of college, my wife and I had some renovations we wanted to get done in our house, so I had told her I was just "exploring" whether I would want to begin training once those expenses were done.

I was being honest when I said that. But the discovery flight hooked me. For those who have been bitten by the bug, it's hard to describe. I often think of a quote attributed to St. Thomas Aquinas:

> *To one who has faith, no explanation is necessary. To one without faith, no explanation is possible.*

Isn't aviation the same way? If you have the "bug" you get it. If you don't, I can't explain it to you. But that won't keep me (or others) from trying. I queried a student pilot Facebook group I belong to on this question of what makes aviation so addictive to

---

[8] Depending on how big and how heavy, that wake can be very dangerous. In December of 2022, a Cirrus SR22 on approach to Knoxville, TN rolled inverted and crashed after flying into the wake of an Airbus A320 about 1.8 miles ahead of it. The pilot was killed. The danger of wake turbulence is mitigated by flying far behind and above the path of a larger aircraft.

so many of us and got many responses that revolved around a few central themes.[9] Freedom comes up a lot. As does the ability to enter a new dimension—the third one to be exact. Many talk about how it reminds them of their own insignificance, seeing the world from a point of view that's more expansive. This resonates with me. Astronauts experience this (presumably to a greater degree) from their orbital vantage points. Some aviators talk about the skill needed to fly and the satisfaction they get from mastering that skill. Still others highlight the community you become a part of when you become a student and, hopefully, private pilot.

These, and so many other things, are elements of what makes flight special to us. I've noticed, though, that most pilots highlight three things that describe what's special about flying.

Observation #1 – Flight uses all you've got

The combination of physical and mental skills necessary to pilot an aircraft are exceptionally unique and drive a sense of satisfaction when perfectly coordinated that no other activity can. While you are flying, your left hand is on the yoke; your right hand is on the throttle, or mixture, or manipulating avionics; your feet are on the rudder pedals keeping the aircraft coordinated. Meanwhile your brain is fully engaged (at least it better be!). What's my altitude? Do I need to adjust power? What will my next radio call be? Am I on the right frequency? How is my speed? Should I put flaps in now or wait a little? Where is that

[9] To find the student pilot Facebook group, just search "Student Pilot Community" in the search bar on Facebook. It's a great group—post a question before a flight and you'll have 20 answers before leaving for the airport.

traffic I just heard on the radio? What did the tower just tell me to do? How much fuel do I have left? There is no other activity I can think of that uses all you have to offer in the pursuit of a successful mission—whether that mission is seeking out a $100 hamburger or flying an organ to a hospital.[10] Being a pilot demands all you can give and rewards you with a sense of accomplishment and freedom unparalleled in any other activity.

Observation #2 – Flight celebrates human ingenuity

We were not meant to fly and so when we do it, we are celebrating what it means to really be human. Ingenuity, creativity, complex problem solving—all these things were needed to conquer the air and are the domain of humans alone. I remember reading an article by a scientist who said that it's easy for us to marvel at the rest of nature and the creatures in it and think, "Wow, how is *that* possible?" Look at a spider's web, or a bee's hive, or a beaver's dam and think of the engineering that goes into making each of those structures. It indeed is incredible to consider what these organisms can do. But the scientist noted that each of those creatures is hardwired to do their "one thing" well, and nothing else. To that point, he closed his commentary by saying, "After all, a spider can't make a 747." Indeed, a spider

---

[10] "Getting a $100 hamburger" is a common aviation community euphemism for flying to get a meal. A burger that might have cost you $8 if you drove to the restaurant suddenly becomes much more expensive if you fly. A recent trip with my wife to a restaurant at an airport in Maine taught me that this euphemism needs updating. By my math, we had a $450 lunch—the salads were about $30 with Diet Cokes and an appetizer. But the rented airplane cost me around $400.

cannot. *But a human can.* While the scientist wasn't making a point about aviation, his choice of an airliner in that statement highlighted for me that among all of humanity's achievements, flight sets us apart from the rest of the animal kingdom more than anything else. Why? Counterintuitively, it's because there *are* animals that fly. And they aren't us. By contrast, there aren't animals that program computers or drive cars or operate jackhammers. In flight alone, it's as if we gave nature, evolution, and physics the finger by showing that even though we don't have wings, light bones, or feathers, we have a brain and an adventurous spirit that cannot be contained. Becoming one with that spirit is what makes flying—as a pilot not a passenger—so special.

Observation #3 – Flight connects you to others

This observation has one obvious angle to it: like other passionate communities, those that share a common interest often bond tightly to each other over that common interest. Obviously, this is not unique to aviation. But what is unique to aviation is the mutual trust that must exist between strangers for such a magnificent thing to be done safely. This point was first made clear to me while reading *West with the Night* a book published in 1942 by Beryl Markham, an amazing woman, pilot, adventurer, and author who noted the following: "A map in the hands of a pilot is a testimony of a man's faith in other men." She is right. And, in modern aviation, her comment is not just true about maps. When you get into that left seat, you are trusting that the engineers who designed the aircraft knew what they were doing. You are trusting that the electronic tools or sectional maps (maps drawn for pilots) you are using are accurate. You are trusting that the controller you are talking to is not asking you to do something that will put you in harm's way. You are trusting that the mechanic who last did maintenance on the aircraft did their job. Please don't get me wrong; you should never blindly trust other players in your aviation adventure. Humans make mistakes. Trust, but to the extent possible, verify. However, there is a limit to what you can verify. In the end you will have to place faith in a complex web of people and systems. And because, unlike stamp collection affinity groups or knitting clubs, your life is literally in the hands of other members of your aviation community, the bond we have to each other is different. At the risk of being overly dramatic, it literally is a bond forged in blood. That may sound macabre, but I believe it's a fundamental reason pilots and others who work in aviation connect with each other. There's an "I've got your back" vibe.

I hadn't codified these observations by the time I landed on Runway 29 on the day of my discovery flight. But they were already tickling my brain as I saw pilots come and go from the tarmac and heard conversations in the flight school offices. I'm sure my immersion in all things aviation early in my life mattered too. But I could just feel it—feel the untouchable, unseeable aura of aviation. And I wanted more of it.

Chapter 2
I'll Take the Beginner Package

After landing, I walked back into the flight school to pay for the flight.

"You also want to buy a logbook?" Bob, an avuncular flight school employee, asked me as I pulled my wallet from my back pocket. I must have looked a little confused.

"To log the first flight of your training," he added, seeming to sense my cluelessness.

What? I thought. *You mean this "counts"*? Indeed, it did. Your logbook will be your record of accomplishment, and that record is used extensively in your PPL journey. Learn how to fill it out early and be as accurate and descriptive as you can. Doing so can be helpful later.

"Sure," I said. I'm confident my tone sounded like that of a little kid who had just been offered an ice cream cone by his grandfather after his mother had told him emphatically, "No sweets before dinner." I felt as if I was getting away with something.

"Do you want to get the other things you'll need too?" he added as he reached under the glass counter to grab a logbook for me.

"Sure, I'd love to," I said.

Wow, Bob was an unbelievable salesman. Either that or he was a Jedi. He'd offer up something while waving his hand as if cleaning a foggy mirror, and I wanted to give him my bank account information.

"What sort of stuff do I need?" I asked, again looking like that little kid, eyes wide at the aviation candy under the glass counter.

"We have a starter package that includes all the books and other things you'll need to become a private pilot."

Bob began pulling items from beneath the glass. Included in the pile were:

- A fuel sump bottle—this is used to drain a small bit of fuel from the airplane into a clear cup to ensure the fuel is the right type and is not contaminated with water or particulates.
- An E6B flight computer—a sort of slide rule we'll discuss shortly.
- A book titled *2021 FAR/AIM*
- A book titled *The Pilot's Handbook of Aeronautical Knowledge.*
- A book titled *The Airplane Flying Handbook.*
- A sectional map of the area I'd be training in—a "sectional," as it's often abbreviated, is an aeronautical map.
- A book titled *The Airman Certification Standards.*

The weight of everything Bob handed me just about matched the gross weight of the plane I'd flown with Jeff and Brittany. None of the books were small. I think it's fair to say that everything I bought that day was critical to my training ... except for one thing. And talking about that one thing will take me into dangerous territory with the PPL community. That one thing is this: the manual E6B flight computer.

And now for some controversy ...

The E6B is a rectangular slide rule device with a circular slider that allows an aviator to make all sorts of useful calculations before and during a flight. It will tell you the distance flown given a ground speed and time. It will tell you what direction to point the nose of the airplane to correct for wind pushing you off course. It can even do unit conversions. It can do these and many, many more functions. Its versatility is amazing when you consider that the original E6B came into existence in the late 1930s thanks to the work of an enterprising navy lieutenant and aviator named Philip Dalton.

I found a picture of Dalton at about 30 years old, and he looks like the sort of guy who would invent a flight computer. With large ears and a pronounced nose, he fits the visual stereotype of a nerd quite well. And I say that with adoration and a sense of comradery. I fancy myself a nerd. My undergraduate degree is in mathematics; I stood in line for hours the week the first iPhone was released to purchase the revolutionary device; I had a Commodore 64 computer shortly after its release in 1983 and learned to program in BASIC on the 64K machine; I had a large collection of *Star Wars* action figures adorning the shelves of my late-teen bedroom. My nerd credentials are stout. So, I consider Dalton a brother from another mother.[11]

---

[11] Continuing the nerd theme—I found the picture of Dalton online at www.sliderulemuseum.com, the home of the International Slide Rule

Dalton was clearly a bright guy. He studied physics at Cornell, got his master's degree in the same subject at Princeton, and then secured a PhD at Harvard, where his thesis was on the topic of artillery fire control, a subject that I suspect has not been studied at Harvard since. Improving on his own original invention, Dalton eventually had a sellable product that he called the "Dalton Aerial Dead Reckoning Slide Rule Model B." Eventually, the device got its current, more famous (at least in the aviation community) name from the Army; E6B is nothing more than a government part number. That said, many ardent fans reject this sterile name, instead calling it the "whiz wheel." It's unclear to me if the name refers to the E6B itself or is some sort of self-congratulatory moniker for the user of the device who might fancy himself or herself a "whiz" for mastering its many functions. The passion about the E6B and the pride demonstrated by those zealots who use it makes me suspect the latter.

It's a testament to Dalton that a device he invented about 30 years after the Wright Brothers first flew was being purchased by me 81 years after his death. I can't think of another instance of such longevity for a manual device that does what an app on a phone can do faster and more accurately.

Yes, the E6B is the Swiss Army Knife of the analog aviation world, performing its myriad functions without a single

---

Museum (ISRM), which touts itself as "the world's largest repository of all things concerning slide rules and other math artifacts." It's really a cool resource, but something tells me that if you have just one "artifact" in your slide rule museum you probably establish yourself as the "world's largest repository of all things concerning slide rules and other math artifacts."

microchip or battery. But there are a million inventions from years ago that we don't hold onto as we have the E6B when more convenient and accurate alternatives come along. Do you use a slide rule to calculate the square footage of your living room? I doubt it. Do you take an abacus to the grocery store? Do you carry leeches with you on business trips in case you find yourself not feeling well?

You can't even make the argument that the E6B teaches you the math of the calculations it performs. It doesn't. Remember, it's a computer, although not an electronic one. It does the work for you.

"But it never runs out of power" is a common retort from those who disagree with me, noting that any alternative to the E6B like a mobile phone app, an electronic flight bag (EFB) application, or a standalone E6B electronic calculator relies on batteries. To that I say, "so what?" I have an electronic E6B app on my iPhone and two iPads I carry when I fly. Are all three of them going to run out of power at the same time? To those who might suggest it's indeed a possibility, I'd point out that my triple redundancy in flight computation is more robust than the redundancy in many aircraft systems that are *much* more critical to safety. Yet we happily hop into our general aviation (GA) planes anyway.

It's also important to note that while I reference EFBs (like ForeFlight or Garmin Pilot) as substitutes for the E6B, those amazing applications don't even require you to do discreet calculations in the same way you might on a manual or electronic E6B. They do all the necessary computations behind the scenes seamlessly, allowing you to do a preflight plan more accurately and safely and, most importantly, make in-flight changes very easily in a much less distracting way with fewer inputs. In fact, since getting my private pilot certificate, I've not used an E6B once.

At the risk of belaboring the point, imagine an alternate universe where Dalton never created the E6B back in the 1930s. In a world where the manual slide rule never existed, anybody who started training sometime in the last decade would have gone through that training using electronic devices to do preflight planning and in-flight calculations. New pilots would never have known the joys of spinning the whiz wheel. Then, one day in our alternate universe, you're sitting in the bathroom, flipping through a Sporty's catalog (don't roll your eyes, you know you do it) and see a full-page ad touting a new, manual flight calculation device, $35 for a metal one and $15 dollars for a plastic one. It's less accurate than the electronic alternatives you've been using. It's hard to read the numbers on it. It will take a while to learn how to use it. It's difficult to use in flight. But good news: you'll have it in case everything you own runs out of power. Would you buy it? No, you wouldn't. The fact is that if Dalton had had access to an electronic computer of the kind we know today when he was at Cornell in the early part of the twentieth century, he would never have designed the E6B as a slide rule. It would have been an app represented by an icon on your phone.

You can tell by the way I started the discussion of the E6B that I knew I was wandering into touchy territory by insinuating that the manual version is a waste of (admittedly not too much) money. Indeed, it's uncomfortable for me to talk about the "E6B question" after just spending so much effort writing about the wonders of the tight-knit aviation community. But with respect to the E6B, it's as if we are a great religion where, like a lot of religions, a schism formed that separated one faction from another and created a world where compromise is just not possible.

I'm not going to do much to bridge that divide here. To those who love the whiz wheel, good for you. Go crazy spinning your wheel, marking your dots, sliding your rule, and reading your answer in three-point font. Take pride in your ability to use it as you might in your ability to operate a rotary phone or a toilet with a flush pull-handle mounted from a box above the commode. But don't try to convince the rest of us, especially those new, impressionable student pilots, you *must* learn how to use it. You don't.

Do you have to be smart?

When you are handed the large books I mentioned earlier and an E6B, you might become intimidated by the idea of putting in the work to become a pilot. It's common to worry that you don't have the intellectual chops to grasp all that you must to earn your PPL. This concern is a valid one. What's strange—and this seems like something that shouldn't be said—is that you don't have to be an *expert* in most of the topics you must know to be a competent and safe pilot. And thank God for that because it's just not possible for any human to know everything. Few people have degrees (or *any* formal training) in all the disciplines powered flight touches: physics, meteorology, engineering, mathematics ... even psychology. Fewer still can apply those disciplines in practical ways as, for example, a mechanic, navigator, or aerospace engineer.

Thank goodness, you might be thinking.

But don't get too comfortable. The fact is you *do* have to have some meaningful intellectual horsepower to become a pilot. There's just no way to grasp the concepts a pilot is expected to know at some level without an inclination—and ability—to learn. I grew up an aviation fan and, throughout my life, have read on the topic. My high school offered an FAA ground school course, and so at 17 I had learned what I needed to learn to pass the FAA written exam for a PPL (while I did pass the exam, I didn't go on to get my license at that time since, shortly after the exam, I shipped off to college). I spent two years as an aerospace engineering major in college before switching to mathematics. I built and flew remote control aircraft, where you get an appreciation for aerodynamic forces and how they relate to control of an airplane. I did all these things and more and *still* had to spend a ton of time studying to get grounded in a multitude of topics.

However, you don't need to have scored in the 99th percentile on a college admissions exam, have a bachelor's degree in anything (much less engineering), and be a member of the Mensa Society to become a pilot. That's not what I'm saying. What I *am* saying is that you need to have some fundamental mental skills to succeed in your endeavor to get your PPL. What are those things and why are they important?

- Intellectual curiosity—you have a desire to understand things you don't, no matter how esoteric they may seem.
    - Why? It's probably obvious, but if you don't have that curiosity, it will be hard to crack a book and stay committed to what you must learn to pass your FAA written and practical tests on your way to becoming a safe and proficient pilot.

- Broad interests—you have curiosity across a large range of unconnected topics.
    - Why? Because becoming a pilot will require you to understand how a piston engine operates, how weather systems behave, how a compass works (and doesn't), and a thousand other things. An interest—indeed a joy—in jumping from one topic to another is a necessity on your PPL journey.
- Creativity—you can visualize things that occur in three dimensions on very large and very small scales.
    - Why? This one is probably a bit surprising. When we think of creativity, we usually think of writing a novel or painting a fresco. But there is another form of this skill, and it entails the ability to read the description of a phenomenon and accurately picture that phenomenon in your imagination. You certainly can get help with this; for example, there are plenty of animations that show how air moving over a wing creates lift or how a weather front moves across a continent. But even so, you need to be able to interpret such images and use them to help you understand why an aircraft behaves the way it does.
- Understanding dependency—you can "picture" and understand how one action or force might influence others.
    - Why? Flying an airplane is all about trading off one thing for another. You trade altitude for airspeed. You trade bank angle for vertical

> lift. You trade endurance for weight. Flight is multidimensional not just in the sense that it occurs in three dimensions, but also in the sense that at any given moment you must understand how doing one thing will affect another.

So, while becoming a pilot doesn't require you to be a genius, *it does require you to have traits that make you want to become a genius,* at least with respect to all things aviation. And these traits are very, very rare in the general population. That's not arrogant to say; it's just a fact. If all humans *did* have these traits, we'd have cured cancer and finally have flying cars in every driveway.

Be ready to read ... a *lot*

I mentioned that in the large bag of items Bob handed to me at the flight school as part of my student pilot package there were several publications. Whether you do what I did and buy them in one go or piecemeal it, you'll need to have each of the items. While all the publications I listed earlier in the chapter are available for free as PDF files from the FAA website, I'd recommend purchasing a hard copy of them from your school or, if not available there, any of the online retailers that sell them. It's nice to have something you can highlight and scribble in to emphasize important points during your studying. But have the PDF files at your fingertips too. One advantage the electronic versions have over the physical (besides the weight and size of the books) is that the PDF files are searchable, so it can significantly speed up looking up a topic.

If studying in the United States, the densest of the books you'll end up referencing is the *FAR/AIM*; it alone is over 1,200 pages. It's actually two books combined into one. The first book, the Federal Aviation Regulations (FARs), can be thought of as the legal do's and don'ts of general aviation. It contains Title 14 of the *Code of Federal Regulations* (CFRs), called "Aeronautics and Space" and is the body of work that, appropriately, covers aviation. Essentially, it's the rules the FAA writes under authorization from Congress. Those rules carry the force of law, so take them seriously. Title 14 is just one of 50 titles in the CFR. Other titles include "The President" (Title 3), "Employee's Benefits" (Title 20), and "Indians" (Title 25).

The broad areas the CFR Titles cover make a lot of sense. But while scrolling down the list of Titles, I noticed one of them, Title 35, was called "Reserved." As you might expect, Title 35 has no regulations associated with it; it's essentially a blank chapter. That begs the question, why is there this random gap in the fifty titles in the *Code of Federal Regulations*? Answer: it turns out that Title 35 used to contain regulations governing the Panama Canal. After the United States relinquished control of the canal in 1999 to the Panamanians, Title 35's original purpose became superfluous.

This is interesting enough, but in my never-ending quest to find odd connections between seemingly disparate topics, I wanted to see if there were any interesting connections between aviation and the Panama Canal. Turns out there's a *very* interesting one.

How are a canal, flying, and Michelin stars related?

In a wonderful *Smithsonian* magazine article by Roy Mize from 2019, a story is told of Charles Field, editor-in-chief of *Sunset* magazine, who published a story written by a guy named Riley Scott called "Can the Panama Canal be Destroyed from the

Air?" The article was published in 1914, just eleven years after the Wrights' first flight. But aviation was exploding, particularly in California where Field lived, and so he took an interest in the new technology. So much of an interest, in fact, that he became involved in a movement in the aviation community to convince the U.S. government of the lethality of aircraft in war-fighting scenarios.

The article itself might have been provocative enough, but the thing that really caught the attention of readers (and the U.S. government) was the photographs that accompanied it. Scott, the author of the article, was not writing about the canal, per se, but about a flight over the canal by an aviator named Robert Fowler who had as a passenger with him, a photographer named Ray Duhem. Duhem had a camera on the flight and took pictures of defenses being built around the canal. The U.S. government didn't take kindly to this and had all four men—the editor, writer, pilot, and photographer—charged with spying. Field argued that he was just trying to "stimulate interest in a larger appropriation for aerial defense in the United States." And, in any case, laws around photography of military installations were specific in their reference to pictures taken from land or sea, not the air. Ultimately, a grand jury seemed to sense Field's passion and honesty and decided the government's case was weak; the charges were dropped.

It's surprising that the government initially pursued the men so vigorously since Riley Scott, the writer of the piece, was a West Point graduate and winner of the 1912 Michelin bombing competition.

The what? you may be asking.

Yes, the French tire maker sponsored a bombing competition. Michelin had been a sponsor of all sorts of aerospace contests and, in 1912, proposed to the Aéro Club de France a contest to see who could most accurately drop a bomb from an airplane. The company's namesake and cofounder, André Michelin, had become concerned by the (prophetic) writings of a French military man who warned of a future conflict where Paris might be bombed by the Germans from the air. Michelin figured he might as well try to persuade France to develop its own bombing capability. You know … just in case. Riley, using a bomb sight of his own design and certainly one of the first made, won the bombing competition.[12]

Anyway, back to the *Federal Aviation Regulations* book. As you might expect by my description of it, the tome is a mind-numbing series of chapters, subsections, and paragraphs that, in the sort of language only lawyers can love, lay out the rules of the (airborne) road for general aviation pilots. Within the regulation there are two parts you'll be asked to pay attention to: Part 61 and Part 91. Part 61 is titled "Certification: Pilots, Flight Instructors, and Ground Instructors." Part 91 is titled "General Operating and Flight Rules." As somebody more creative than me once said: Part 61 tells you how to get your license and Part 91 tells you all the ways you can lose it. I think that's a very accurate description of these parts.

---

[12] You may be familiar with Michelin stars being awarded to fine dining establishments. The genesis of this predates the bombing competition that Michelin sponsored in 1912. At the turn of the century, with few cars on the roads, Michelin hoped to promote auto travel (and thus sell more tires) and so began publishing travel guides, which also noted the location and services of repair shops, hotels, taverns, etc. Eventually, restaurants were included and, in 1926, those restaurants were scored using the three-star system that is known today.

Don't be too intimidated by the size of the book. There's no arguing that there's a lot of pages and unreasonably tiny type filling them. But you won't be expected to know everything in it by memory. Some things, yes. But not all things. The items you don't commit to memory you should be able to find in short order. So, spending some time going through the high-level structure of the book and, as most pilots-to-be do, "tabbing" relevant sections for quick references to common statutes, is recommended. During the oral exam, which will likely be administered by a designated pilot examiner (DPE) in the final step before receiving your certificate, it's likely he or she will ask you a question you will have to look up. That's okay. Just be prepared enough that any venturing into the pages is a quick surgical hunt, not a meandering walk through the savanna hoping you stumble upon your prey.

The *Aeronautical Information Manual* (*AIM*) is easier reading than the *FARs*. Admittedly, that's a low bar, like saying that hiking Everest is a little easier than hiking K2. The *AIM* is what its title suggests: an information manual about navigating the skies. In fact, its subtitle is "Official Guide to Basic Flight Information and ATC Procedures." It lives up to this subtitle, with a ton of great information that reads like a sort of "how-to" book for flying an airplane in the United States. If you are in a different country, you will have an aeronautical information manual for your airspace (the United Kingdom, for example, has its own).

While the information in the *AIM* is very important—and in fact includes more practical information for flying than the *FARs*—I found that I referenced it less. This is because much of what you'll need to know in the *AIM* you will learn by doing, by crawling into a cockpit and flying. And, in any case, there are many esoteric things in the *AIM* that you likely don't have to have a deep (or any) knowledge about. For example, there is a section regarding interception procedures, referring to the proper method for national defense aircraft to intercept other aircraft. There's another that details where "FAA Sponsored Explosives Detection Dog/Handler Teams" are located.

Still, get familiar with the *FARs/AIM* by cracking it open regularly from the time you decide you want to get your PPL. At first, doing so will be like visiting a boring uncle who uses big words to impress you without really saying much. But over time, you'll view your uncle with affection and absorb his nuggets of wisdom without even realizing you are. Then, one day, you'll be in a "conversation" with an FAA examiner on the last step before getting your private pilot certificate, and all that knowledge your boring uncle dropped in visit after visit will come back to you in a rush, allowing you to answer the question, "How long is a Class 3 medical certificate for someone over 50 good for?"

Are you healthy?

Speaking of medical certificates, you probably aren't surprised that your health is something the FAA cares about. If they're going to permit you to put a hand on a yoke or stick and another on a throttle and hurl a piece of aluminum and steel occupied by human beings into the air, they're going to want to have some confidence you aren't going to keel over in the middle of your flight or do something stupid because of your mental state.

You are therefore required to get a medical blessing from an approved FAA Aeromedical Examiner. This is something you should do soon after deciding to get your PPL. As the FAA's own website says matter-of-factly, "We suggest you get your medical certificate before beginning flight training. This will alert you to any condition that would prevent you from becoming a pilot before you pay for lessons." Prudent advice. You'd hate to spend many thousands of dollars before you realized that it might have been better to bet that money on the ponies because you're unable to fly as a pilot in command.

There are three types of certificates: Class 1, 2, and 3. You need to have at least one of them to become a private pilot, but a Class 3, which is less onerous than the Class 1 and 2, is sufficient. You must have the certificate in hand before you fly an airplane solo, but you don't need one to fly with a CFI. This is why many student pilots can get lulled into starting their flight training journey and be well into it before realizing they may not be able to get a medical certificate. (While you should get your medical certificate early, don't get it before you've taken a couple of lessons. It can cost a couple of hundred dollars for the medical exam and, as I noted earlier, some people realize after experiencing what flying is really like that their lifelong dream really isn't their cup of tea. So, while getting a medical okay is an important thing to do early, there is such a thing as being *too* early.)

I found getting the certificate straightforward. You need to visit an FAA-authorized medical examiner, of which the FAA says there are over 6,000 in the United States. The FAA website contains a searchable database; your school or instructor may also have some recommendations.

I used the FAA website to find an examiner near Hanscom and made an appointment. When I arrived for my exam, I was thrown for a bit of a loop. I had expected to pull up to a medical office building of some sort but instead found a driveway leading to a mid-century modern home on a heavily wooded lot. I called the examiner to make sure I was in the right place and, sure enough, she emerged from her home, walked to my car, and, after I rolled down my window, asked me to meet her in the garage as if we were doing a drug deal. Indeed, if I hadn't found her on an FAA website, I might have thought I was on my way to becoming a story on a *Dateline* episode. *He just wanted to be 6,000 feet up … instead, he ended up six feet under.*

My worries were unfounded. She was probably in her mid-60s and moved slowly but with purpose, like a grandmother preparing a Thanksgiving meal.

The exam was much like a physical you might get at your family doctor, but you'll be glad to know no latex gloves are involved, and you can keep your pants on (and if you were asked to take them off, you might have an opportunity to be on a "To Catch a Predator" *Dateline* episode).

While she filled out paperwork, I asked her how she got into doing FAA medical exams. She explained that her husband had been a pilot who flew out of Hanscom, right around the corner. When he was in the prime of his flying days, she would fly with him and that led her to become interested in aviation and that, in turn, led her to parlay her medical training into a role as a FAA medical examiner. I got the impression that she didn't practice medicine anymore except in this capacity. It was a semi-retirement gig that likely enabled her to deduct some of her mortgage expense as a business expense, although I wonder if the IRS might get picky if you identify your garage as your exam room. I had to work my way around the bumper of her car when backing up to a line on the cement where I was to stand and read an eye chart.

If you get an annual physical with your primary care physician, and no issues have been found, you should expect the same from your FAA exam. This was the case for me. After my exam was over, I waited in my car while my examiner went back into her Brady Bunch home and printed my medical certificate on a standard 8.5x11 piece of paper. In the "limitations" section, there was only one note: "Must have available glasses for near vision."

While the limitation specifying the use of glasses was benign, I must admit a bit of post-traumatic stress came to me on reading it. It took me back to the late '80s when I attended the University of Notre Dame on an Air Force ROTC (Reserve Officer Training Corps) scholarship. The scholarship was a vehicle for the Air Force to pay my tuition and, in return, I'd owe them four to six years as an officer upon graduation depending on my career field. I hoped to become a fighter pilot but knew this was unlikely because, from a very young age, I wore big, thick glasses that made it look like I was wearing glass dinner plates over my eyes. I tried to weasel my way into navigator training where the requirements were less stringent, but my astigmatism was too severe to meet even these lower standards.

It was all hugely disappointing to me. It wasn't just disappointing; it was maddening because it was exceptionally unfair. There was a common myth at the time (and it persists today) that perfect eyes are needed to be a fighter pilot in the American military. But this is untrue in its strictest sense and has been for a very long time. During my time in ROTC, the requirements to *enter* pilot training indeed stipulated perfect vision. But once you *were* a pilot, you could have significant deterioration in your eyesight and would be allowed to stay in the cockpit if corrective lenses got you back to 20/20 vision. Think about that—the requirement was necessary to get the job but completely unnecessary to do the job. In any other context it would be unlawful discrimination. But for some reason, the military was allowed to do it.

In the Air Force's own archives there is evidence of the folly of this requirement. I found a study published in 2017 done by two Air Force officers looking into how the astigmatism of applicants who were granted waivers to enter pilot training progressed over a five-year period (such waivers were very, very rare in my day). Those granted waivers had astigmatism at a level that would require the use of glasses to see clearly. While the focus of the study wasn't how their poor eyesight affected pilot performance, if at all, there was this telling observation buried in the discussion: "No subjects developed corneal ectasia [the progressive condition they were researching] and no subjects were removed from flight status for reasons related to their eyes or vision…. On the basis of Safety Center data, no subjects were involved in a mishap where eyes or vision was a contributing cause." Sure, it was one relatively small study. But as I noted at the beginning of this discussion, it doesn't really matter. The only evidence you need to understand the arbitrariness of the perfect eyesight rule is this: once the military invested training dollars in you, they let you put on the Coke bottle glasses and keep flying.

Fortunately, the FAA is more forgiving, probably because they don't worry about me having to shoot down an enemy aircraft. In fact, color blindness is probably more of an issue than visual acuity. The latter can be corrected in most cases with glasses or surgery, the former cannot. And being able to see colors is an important part of being a pilot. But even color blindness does not mean you can't fly at all. Depending on severity, you may simply have a limitation (no night flying for example) stipulated on your certificate. Oh, and there's good news for our female readers—color blindness is much rarer in women than men. Only one in 200 women are affected; one in 12 men are.

Always think happy thoughts

FAA concern about physiological and mental issues getting pilots killed has existed for a long time. What's interesting is that medical issues *not* related to mental health are very rarely the cause of aviation accidents. Specific data on this with respect to general aviation is hard to find. But in any list of the top ten causes of general aviation fatal accidents, I couldn't find one that listed "medical incapacitation" or the like as a distinct category.

There is better data on this in the world of commercial pilots. In a study published in 2004 by the U.S. Department of Transportation's Office of Aerospace Medicine titled *In-Flight Medical Incapacitation and Impairment of U.S. Airline Pilots: 1993 to 1998*, the authors found aircrew members became incapacitated or impaired due to a medical event a total of 50 times during the five-year period. Given that U.S. airlines flew 85,732,000 hours of revenue-producing flights during that time, these events were exceedingly rare. None of them resulted in a fatal accident (and only two resulted in an accident at all). Obviously, one reason they are rare is because the FAA does a good job of keeping high-risk individuals out of the cockpit. And when events do occur, the outcome for the passengers is generally good since most commercial flights have two pilots on board. (The only instance I'm aware of where the entire aircrew became incapacitated is in the movie *Airplane!* Don't eat the fish!)

That's not to say things didn't get dicey at times. One incident in the report was described this way:

A 45-year-old B-737 first officer experiencing an alcohol withdrawal seizure suddenly screamed, extended his arms up rigidly, pushed full right rudder, and slumped over the yoke during an approach. The aircraft descended to 1,000 feet above ground level in an uncoordinated turn to 25 degrees angle of bank before flight attendants could pull the first officer off the controls, allowing the captain to recover the airplane. "Mayday" calls were made, and the captain executed a missed approach before making a successful landing.

"Colorful" stories like this are not reserved for the commercial aviation world. In fact, given the less stringent medical and training standards required to fly as a private pilot in the general aviation community, there are more cases and more bizarre ones. Many of these are unrelated to physical well-being and instead have to do with mental issues.

In 1969, two physicians working for the FAA wrote a report titled: "Medical Factors in U.S. General Aviation Accidents." I dove into the report expecting I'd need an espresso to keep me awake. But the authors, doctors Siegel and Mohler, must have had a bit of marketing expertise because the first case they delve into reads like a scene from a James Bond film. It's a tale that has everything: titillation, daring, and anguish. And it's all wrapped up in the staccato rhythm of short sentences that remind you it's a government-funded report. I quote it at length here:

The very experienced pilot of a Stearman 75 made a series of low-level flight maneuvers over a river in Hawaii, including a flight under a bridge followed by an abrupt pull-up, in July of 1967. Afterwards he flew to a section of beach and waved at a girl in a Bikini who waved back....

Let me interrupt our tale for a moment. If I hadn't told you when this report was written, you probably would have known it was from a different era. I'm not sure the FAA brass would approve today of language that references a "girl" much less the fact that she was wearing a bikini.[13] Why in the world is it relevant that the "girl" was in a Bikini? And why was it relevant that the pilot waved at her, and she waved back? Continuing ….

> *The pilot landed on the beach nearby, introduced himself to the girl, then borrowed her telephone, calling several bars to locate his girl friend [sic]. While doing so he drank a beer. The pilot then departed saying that if his newly met acquaintance keeps wearing her Bikini, he was coming back. He also said, "Always think happy thoughts."*

Okay, now it's just getting creepy.

> *On becoming airborne he executed two loops near ground level, and on the second loop impacted the ground with fatal consequences. It developed that the pilot had recently separated from his wife and in past years had been known at times as a very heavy drinker. The*

---

[13] You may notice that the word "Bikini" is written with a capital letter in the report. The term was once a trademarked brand name before it became generically used for any two-piece bathing suit. The creator of the modern Bikini was a French man named Louis Réard who, according to a *Time* magazine web feature, finished his design in 1946, just four days after the U.S. had detonated its first peacetime nuclear weapon on the Bikini Atoll in the Marshall Islands of the Pacific. Réard hoped his new suit would be as explosive as the U.S. nuke, thus the name. Ah … the good 'ole days when fashion got inspiration from the detonation of nuclear weapons.

pilot had also told persons for some time that he was going to fly under the bridge if it were the last thing he ever did.

The report goes on, after conveying this story, to note that "the use of an aircraft for expressing sexual frustrations … [is] documented."

I'm making light here of the presentation of the case, not the case itself, which is tragic to be sure. And I'm highlighting it to emphasize the point that mental health matters as much (and maybe more) than physical health when acting as a pilot in command. While we think of mental illness as a discreet mental disorder (e.g., depression, bipolar disorder, etc.), we'll learn throughout this book that virtually all fatal accidents are, in the end, the result of a form of mental impairment. And that's because almost all accidents are the result of poor decision-making. That fact may have you wondering as you begin your training, "Just how safe is this thing I'm getting ready to do?" That's a tougher question to answer than you might think.

Chapter 3
How Safe Is This?

Shortly after beginning my PPL training, the worries started to accumulate in family and friends.

"Aren't you scared?" was a question I got at a social gathering one night shortly after I'd let people know that I'd done my discovery flight and was now committed to the hours of training it would take for me to become a private pilot.

"It's safer than driving a car!" somebody else in the circle of conversation responded before I had a chance to.

But is that true?

It *is* true that commercial air travel is immensely safer than driving a car. How much safer depends on your data source. The numbers with respect to airline safety are surprisingly all over the map. This is partly because there's more than one way to approximate relative safety. In the interest of picking a single, reliable source on airline safety, though, I refer to data provided in a paper written by MIT professor Arnold Barnett.

Barnett published research in *Transportation Science* in 2010 that looked at the likelihood of death from commercial air travel, but with a bit of a twist. He broke the world of commercial aviation into three groups: first-world countries (e.g., U.S., Canada, Japan), developing world nations (e.g., Singapore, South

Korea), and newly industrialized nations (e.g., China, Brazil). He found that the risk of dying in a commercial airline crash *overall* was one in 3,000,000 per flight. However, when the data was looked at based on the country groupings mentioned earlier, the numbers are very different. The first-world countries had a death risk per flight of one in 14,000,000. In the developing world the number was one in 2,000,000. And in the newly industrialized nations the mortality risk was one in 800,000.

But even if we take those last numbers and compare them to your chances of being killed in a car accident, you'd still rather be flying from Cameroon to Liberia than driving from Peoria to Chicago. That's because your lifetime risk of dying in a car accident (in the United States) is one in 107 according to an analysis by the National Safety Council. You'll note that this number is a lifetime risk, while the aviation numbers I quoted in the previous paragraph are your chances of dying on any given flight. So, we are doing a bit of apples to oranges comparison here. But suffice it to say that no matter how you slice it, driving is considerably more dangerous than flying on a commercial airliner.

But you're not flying commercial

But what about general aviation? Many don't realize that it's distinctly unlike commercial aviation. First, there's the level of training. When going through your pilot training it may seem like you're being put through the ringer, yet it's incredible to think how *little* is required of you to get a PPL in the United States. You need to log just 40 hours in the cockpit of an airplane before getting permission to operate an aircraft with passengers (this is the minimum time; most will take longer). Think about that. Spend the equivalent of one standard work week in the left seat of an airplane, and the FAA might sign you off to act as the pilot in command of an aircraft. Airline pilots, on the other hand, go through extensive training and—perhaps more importantly—extensive *recurrent* training. And between episodes of that training, they are flying a lot, making it easier to stay current. Second, there's the fact that most commercial aircraft have two pilots in the cockpit.[14] That's a lot of redundancy. They check each other's work and help lighten the workload on any one individual. You own everything when you fly as a GA pilot; in most cases there's nobody there to tell you that you are making a dumb decision or missed a checklist item. Third, the technology in many commercial airliners is beyond that you'd find in most general aviation aircraft. Onboard radar to see weather in real time, sophisticated anti-collision systems to keep aircraft from running into each other, and two (or more) very powerful engines to mitigate the loss of power in one … all these things and many, many more increase safety substantially. Finally, there is the ever-present understanding among the flight crew that doing something stupid—even if it might not kill you—could cost you your job. If you do something stupid in the cockpit of your private plane, nobody may ever know. Get away with that stupid thing and you may do it again until … one day …

No, general aviation is *not* like commercial aviation. And, not surprisingly, the statistics bear this out. I found the raw data from the FAA on general aviation fatalities between 2001 and 2020

---

[14] I say "most" here because some smaller air carriers operate single-pilot aircraft. For example, in the part of the world where I live, New England, Cape Air flies single-pilot aircraft between destinations around the Northeast. In fact, if you fly with them, you may, as a passenger, get to sit in the copilot seat for weight and balance reasons.

and used that to do some analysis. I focused on the last eight years of data (2012 to 2019). I did this for three reasons. First, data is inexplicably missing for 2011. Second, safety increased during the full 20-year period, so I wanted to account for the better numbers over the later years' data set. Third, I wanted to compare the aviation numbers to driving numbers and, as of the writing of this paragraph, the latest reported year on driving fatalities from the National Highway Traffic Safety Administration (NHTSA) is 2019.

During this time frame from 2012 to 2019, there was an average of 18.4 general aviation fatalities per million flight hours. What was it for commercial air travel inside the United States? *Very* small: 0.095 fatalities per million flight hours. General aviation flying is about 194 times more dangerous than commercial flying by this metric. It's not surprising when you consider how many flight hours commercial operators rack up and that during the period we are considering, there were no fatalities at all in five of the eight years.

That differential between GA risk and commercial risk is shocking. But commercial air travel is *so* safe, comparing any other cause of death to it looks shocking. After all, that 0.095 number is due to a total of 13 deaths in the eight-year period. In the United States in 2019 *alone* there were 19,000 homicides, 33,000 motor vehicle deaths, 39,000 deaths from falls, and 20 fatalities from lightning strikes.

Comparing GA flying to commercial flying, then, seems unfair. How then does GA flying look compared to an activity virtually all of us partake in: driving? It's very tricky to make a one-to-one comparison. As you've noticed, fatalities in aviation are based on flight hours. And even that is an educated guess when it comes to the GA world since there is no formal tracking mechanism of hours flown.[15] Driving fatalities, by contrast, are measured in hundreds of millions of miles driven. During our same reference period of 2012 to 2019, there were 1.1 fatalities per 100 million vehicle miles driven in the United States. That may seem like a low number, but keep in mind that the total miles driven in the U.S. every year is about 3.2 *trillion*, so the total fatalities figure is quite large: around 32,000 deaths a year during the period.

One method some[16] have used to try and make a comparison between driving and GA flying is to convert hours of flying into miles flown. If we assume an average ground speed of 140 mph (around 120 knots) for general aviation aircraft, that means that 1,000,000 flight hours equates to 140,000,000 statute miles traveled. Our 18.4 fatalities per million flight hours would therefore be the same as 18.4 fatalities per 140 million *statute miles traveled*, or 13.1 fatalities per 100 million statute miles traveled. This is obviously not a heartening number; it's about 13 times higher than the car crash numbers which, as a reminder, were 1.1 fatalities per 100 million statute miles traveled. But again, we are assuming that the "deaths per 100 million miles flown" number we backed into for aviation is a good way to do a comparison. It may not be … more on this in a minute.

What about motorcycle riding? According to the NHTSA, all you easy riders out there face a fatality rate of about 25 deaths

---

[15] I suspect that with new aircraft tracking systems like ADSB (which I'll discuss later), it may be possible for the FAA to get much more accurate data on hours flown by GA aircraft in the years ahead.

[16] The most well-known book on safety in general aviation is *The Killing Zone: How and Why Pilots Die* by Paul Craig. While I use a different data set here, I use logic similar to Craig's to try and compare flying to driving.

per 100 million miles traveled. This is nearly two times greater than the number we just calculated for GA flying and nearly 25 times greater than driving a car.

But some good news ...

We are therefore left with numbers that tell us that the risk of GA flying is about halfway between the risk of driving a car and the risk of driving a motorcycle.

However, there's a really important point to consider when thinking about these comparisons. And it has to do with this: you could die in a car or on a motorcycle due to no fault of your own. You could be happily driving through a green light with your dog in the passenger seat when a driver who has the red light is distracted while texting his girlfriend and plows through the intersection, sending you across the rainbow bridge with your dog. You're not in complete control of whether you have a safe drive or motorcycle ride.

But I'd argue—and I think the data supports—that you *are* almost always in control of whether you have a safe flight.

How can I say this with such confidence? You probably already know: most aviation fatalities are the fault of the pilot. I looked through data that the Aircraft Owners and Pilots Association (AOPA) puts together in something called the *Joseph T. Nall Report.* Joseph T. Nall was a member of the National Transportation Safety Board (NTSB), the organization responsible for investigating aircraft crashes. In a stroke of sad irony, Mr. Nall was killed in 1989 while flying as a passenger in a small plane in Venezuela believed to have been chartered for a sightseeing flight. The report now named after him includes detailed data on the number of accidents each year, how many were fatal, and the cause of those accidents. While the numbers bounce around a bit year to year, somewhere around 75% of fatal crashes are classified as "pilot-related." The report defines these crashes as "accidents arising from the improper actions or inactions of the pilot." The "mechanical/maintenance" category is often below 10% of fatal cases in a given year, and the balance of cases are classified as "other/unknown."

Let's assume, conservatively, that 70% of all fatal crashes over the eight years we discussed earlier were due to pilot behavior. That would mean that of our 13.1 fatalities per 100 million miles flown, only about four of them would be due to an act of God—an engine coming apart at a point where there was no chance of recovery, a wing falling off due to a manufacturing defect, a bird strike ripping off the tail. While this number is still about three to four times greater than the fatality rate in automobiles, it gets close enough to be comparable. And it's well below the risk of riding a motorcycle.

I want to emphasize one last time that these numbers only point us in the direction of understanding risk; they are not precise in any way and, in fact, have some big flaws. For example, consider that most fatal GA accidents occur during takeoff and landing. An NTSB study found that between 2013 and 2018, 23.6% of GA accidents occurred during takeoff and 40.2% during landing. With that information in mind, imagine two pilots flying from the same airport with the same skills, the same decision-making ability, and the same appetite for risk. They are essentially identical twins flying Cessna 172s *in exactly the same way.* Their airplanes will not fail mechanically; they are 100% reliable. Now imagine that one of our pilots takes off, does just one lap around the pattern, and then lands, flying a total of five miles. The other takes off, flies 100 miles away from the airport, turns around and comes back, and lands in the exact same conditions (same winds, lighting, cloud cover, runway condition, etc.) as the first pilot. Each of our imaginary twins had one takeoff and one landing under the same conditions. Therefore, they faced substantially the same risks. Yet one flew 40 times more miles and about two hours longer than the other. Thus, measuring accident rates or fatalities by miles flown—or even hours flown—is convenient, but not precise in any real way.

But it's all we've got. And it does allow us to understand relative risk in a directional nature. Remember that in the eight-year period we discussed earlier, there were a *total* of 13 deaths due to commercial airliner crashes. In 2020 alone, general aviation registered 332 fatalities. And GA pilots flew far fewer hours and miles than their airline counterparts.

No matter how you look at the data, the key takeaway must be this: GA flying has risks, but they are manageable. And managing that risk depends almost entirely on you and your decision-making.

Chapter 4
The Forces of Flight

I began this book by talking about a British aristocrat and now, I introduce another: Sir George Cayley. Cayley lived a generation before Haldane. He died in 1857, the year after Haldane was born. If you've never heard of him, you're in good company. But while his name may be unfamiliar, the output of his efforts is not. It was Cayley who first suggested that there are four forces acting on an aircraft in flight: lift, weight, thrust, and drag. If you asked student pilots what image they first remember from their study to become a private pilot, I suspect the iconic image of a high-wing trainer with arrows pointing off its top, bottom, nose, and tail labeled with their respective forces would top the list.

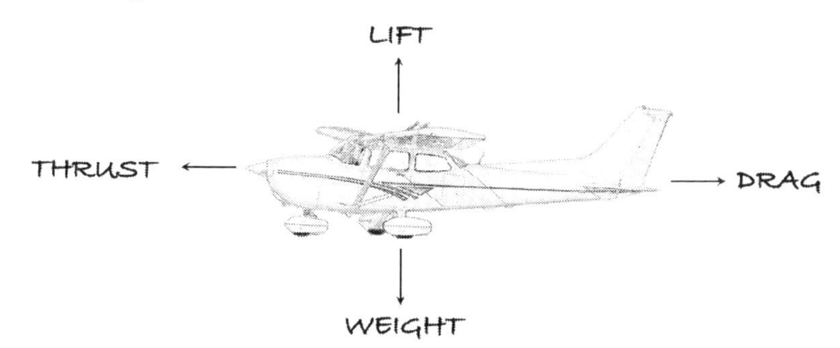

Lest you think coming up with this framework for the forces acting on a plane in flight was obvious, note that in Cayley's days

(days before any powered flight existed), people were still convinced that the way to get airborne was to mimic a bird. In fact, there was a lot of time, effort, and money invested in the development of ornithopters—contraptions that had flapping wings. These machines attempted to imitate the wings of birds and insects by using a single surface to both create lift and to provide forward motion. Cayley was the first person to suggest that maybe both lift and thrust didn't have to come from the wing. With an incredible sense of what the future would bring well before it was obvious, he envisioned what the airplane would eventually look like: a fixed-wing craft (no flapping) with an engine providing thrust.

While engine technology didn't allow him to prove his insight while he was alive, that didn't keep Cayley from doing more than just thinking big thoughts. He was an experimenter and designer and is credited with building the first airplane—a glider. In 1849, one of his designs carried the first human (the 10-year-old son of a servant) into the air in a heavier-than-air machine.

Cayley probably didn't appreciate during his own lifetime his genius when it came to identifying the four forces. I wonder what he'd think if he opened Chapter 5 of the FAA's *Pilot's Handbook of Aeronautical Knowledge* (*PHAK*) and saw several pages dedicated to his insightful model of the aerodynamic forces acting on an airplane. Included in those pages are graphs and illustrations that can be intimidating. Don't worry though; it's not necessary to understand the details of every image embedded in the text. But you will have to understand the underlying principles related to these forces and the energy associated with them. Because ultimately a pilot is a manager of an airplane's energy. I'll go into more detail later about what I mean by energy management, but that discussion will be more fruitful if we first establish a common understanding of the forces acting on an aircraft in flight and how an aircraft moves about various axes due to the control inputs of you, the pilot.

The forces defined

For most, the definition of the four forces Cayley first identified is intuitive. Unlike many aspects of the aerodynamics of flight, even the novice can understand that there must be something that pulls the airplane up. That "something" is called lift and is what makes an airplane unique. In the next chapter, we'll talk in detail about what causes lift, but suffice it to say that if our plane is to get off the ground, the force of lift must overcome the force of weight.

To generate lift, the wing needs air moving over it. And to provide that moving air, we use an engine to develop thrust. Thrust provides forward motion, the forward motion provides moving air, and the moving air provides lift when it hits the wing.

But the thrust must overcome the resistance that very same air creates. That resistance is known as drag. Drag pulls against the plane as thrust pulls it forward. Drag isn't constant. It varies depending on the configuration and speed of the airplane.[17] This is important to understand, and you'll quickly realize that if drag *was* constant, it would be a lot easier to fly an airplane.

Finally, weight is the force of gravity acting on the mass of the airplane.

---

[17] "Configuration" here refers to the position of elements of an aircraft during different phases of flight. For example, when an airliner is configured for landing, the landing gear will be down and flaps will be deployed. Both things expose more surface area of the airplane to the oncoming air, therefore increasing the amount of drag it experiences.

When an airplane is in straight and level flight at constant speed, the opposing forces sum to zero; they exactly cancel each other out. This too makes sense intuitively. If weight is balanced with lift, the airplane will "hang" in the air; it won't climb or descend. If thrust is equal to drag, the plane will be at a steady speed; it's not accelerating or decelerating. See, it's all quite simple.

Roll, Pitch, and Yaw

While Cayley's forces are pulling, pushing, and lifting your plane, you will control it in three dimensions using roll, pitch, and yaw. Each of these movement types is defined as rotation about one of three axes. The longitudinal axis runs from nose to tail and is the axis around which roll happens. The lateral axis runs

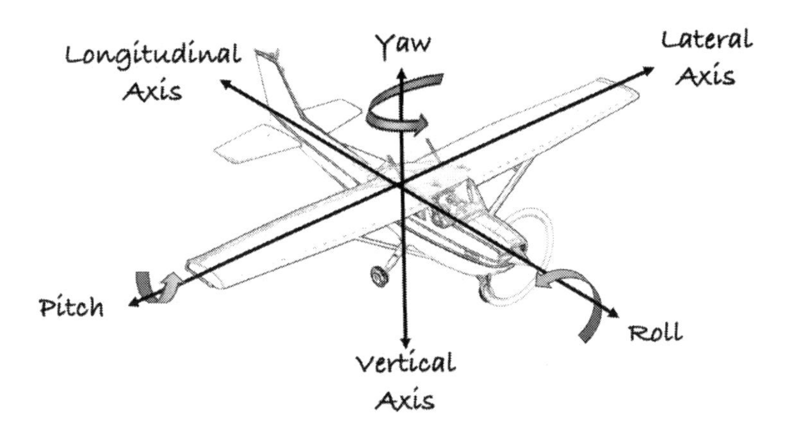

from wingtip to wingtip and is the axis about which pitch happens. And the vertical axis runs straight up and down through the center of the plane and is the axis about which yaw happens. As you might have noticed, the primary control surfaces on your airplane are associated with each of these axes. Roll is controlled by use of the ailerons, pitch by use of the elevator, and yaw by use of the rudder.

Drag and lift

I noted in our earlier discussion of Cayley's forces that drag is not constant and hinted that this is an important thing to understand. In fact, while the four forces are interesting, understanding what happens to drag in relation to lift is where the rubber meets the road—or, more accurately, where the wing meets the air.

If you dug into Chapter 5 of the *PHAK* you probably saw a graph or two that showed the relationship between lift and drag. You'll notice that as lift increases, so does drag. The chapter also shows that we can increase lift (and will therefore increase drag) by increasing the wing's angle relative to the oncoming wind. The faster we go, the lower that angle can be to generate a given amount of lift. The slower we go, the higher the angle needs to be to generate the same amount of lift. If you've ever been to an airshow and seen an airplane make a slow pass while maintaining altitude—i.e., its lift matched its weight—you probably noticed how high the nose was and, therefore, how highly angled the wing was relative to the same airplane making a high-speed pass.

Another example of this lift-versus-angle-of-the-wing-versus-speed interplay is made obvious when you consider what happens during a landing. The nose of the aircraft is raised in the moments before touchdown because if it is to generate enough lift to let itself down gently as it slows, the pilot needs to increase the wing's angle relative to the oncoming wind to continue to generate sufficient lift.

This relationship between speed, lift, and the angle of the wing relative to the oncoming wind (something we will define in the next chapter more formally) is essential to understanding how to control an airplane. In fact, as the *PHAK* points out, "Two major aerodynamic factors from the pilot's viewpoint are lift and

airspeed because they can be controlled readily and accurately."
With that baseline explanation of the forces of flight, let's now focus more fully on lift—what it is and why understanding how it is generated is so important.

Chapter 5
Why Planes Fly

Within a month after my discovery flight, I had flown another four times with Jeff. I liked him even though he could, at times, seem a bit scatterbrained. It didn't help that he had a mop of hair that was perpetually on high alert as if he'd ridden a motorcycle to work without a helmet. He could look like a confused Muppet. But he was undoubtedly a good pilot and teacher.

In early March, during my third lesson, Jeff and I took off from Hanscom and flew about 15 miles to the southwest where we entered the flight school's designated practice area. There, Jeff demonstrated various maneuvers, explaining their relevance to me as a future pilot.

The maneuvers he had me do were all part of the *Airman Certification Standards* (*ACS*). You will recall this was one of the books I bought from the flight school after my discovery flight. It's the "bible" when it comes to what you'll need to know to earn your PPL. As you thumb through it, you'll likely be hit with a shock of intimidation. There are a LOT of things in the *ACS*. It includes things that are best classified as "book learning" as well as things you'll have to demonstrate when at the controls of your airplane. This mirrors the way the day of your checkride (your "final exam") will go—an oral exam on the ground and a flight exam where the FAA examiner will observe your ability to perform prescribed maneuvers.

While all sections of the *ACS* are important, Section IV, titled "Slow Flight and Stalls," is one of the most important. Four pages of requirements are in this section and for good reason. One of the greatest dangers you will face as a pilot is the risk of a stall. The good news? You have virtually complete control over whether your aircraft will enter a stall.

By now, you probably know that a stall in the context of an aircraft has nothing to do with its engine. It's an unfortunate name given the more common use of the word to refer to an engine quitting. It's painful once you know something about aviation and aerodynamics to hear a journalist conflate an aerodynamic stall with a dying engine. Then again, a closer inspection shows that the use of the same word for seemingly unrelated phenomena makes a lot of sense. The word comes from the Middle English word *stallen* which can mean "to come to a standstill." It's the reason the cordoned-off area in a stable where a horse can only stand or lie is called a stall. It's also the reason a similarly cordoned-off area in a public restroom is called a stall. This theme of the word *stall* having to do with the inability to move, then, carries through to aviation. Except in the context of your airplane, a stall isn't about the plane not moving, but about air not moving in the way it should over the wing (which can abruptly lead to the aircraft not flying).

To understand a stall, it's necessary first to understand the structure of a wing and why it creates lift. Fortunately, a wing is simple. For the purposes of our discussion, you only need to know three parts of the wing: the leading edge, trailing edge, and wing chord. The leading edge and trailing edge definitions are probably obvious. The leading edge of the wing is the part that faces the wind and is therefore the first part of the wing to meet the air. The trailing edge is the "back" of the wing. The wing chord is simply the line from the leading edge to the trailing edge.[18]

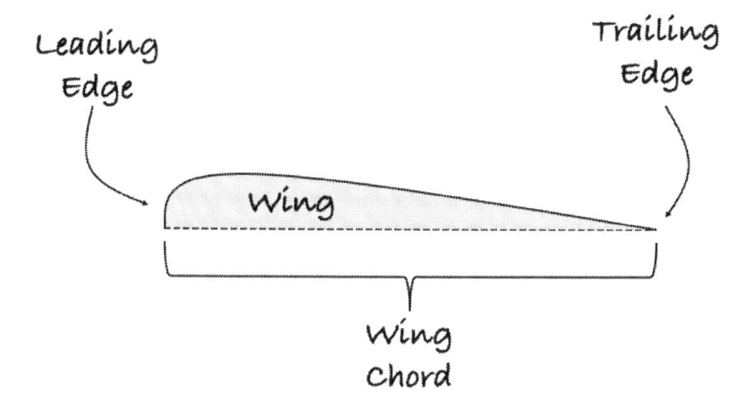

These three terms are all you need to know to discuss more complex topics, like how a wing produces lift. To that point, the *PHAK* discusses what causes lift in Chapter 4, and, like many other explanations, it turns to the work of Daniel Bernoulli. Bernoulli was a Swiss mathematician and physicist who discovered that when a fluid like air increases in speed, pressure decreases. Because the shape of a typical wing (and/or its angle to the air) makes air flow over the top of it faster than the bottom, Bernoulli's discovery comes into play and the pressure on top of the wing drops relative to the pressure on the bottom. This

---

[18] To keep things simple, I've shown a wing cross section here where the bottom is flat and therefore defines the wing chord. Many wings though are curved on the top *and* the bottom. In this case, the chord is still defined in the same way; it runs from the leading edge to the trailing edge but would show up in our cross section running through the middle of the wing.

differential in pressure effectively sucks the plane into the air; you are airborne.

But I had always wondered: Why does the air *have* to flow faster over the top? Many years ago, I saw an animation that attempted to explain. It showed a group of air molecules hanging out together when, suddenly, a wing—curved on the top and flat on the bottom—pushed itself through them. One parcel of molecules was forced above the wing and one parcel below. The voice-over of the animation explained that those two parcels of air will want to "reconnect" by crossing the bottom and top surface of the wing in the same amount of time to arrive at the trailing edge of the wing simultaneously. But since the top parcel of air must travel farther than the bottom parcel because the curved route over the top of the wing is longer than the route over the bottom, it must move faster to arrive at the same point at the same time (something known as the "equal transit time theory"). This more rapid movement is where Bernoulli's principle comes into play; faster-flowing air means lower pressure.

I had always taken this explanation as an article of faith. But tickling my brain for years has been this question: *Why* do our two parcels of air molecules *have* to get back together after the wing pushes them apart? In my time studying physics while in my aerospace engineering program, I never recalled learning of some law that said parcels of air needed to be connected to each other, never to go their own way—of course, I dropped aero engineering because my grades were crummy, so maybe I slept through that part? But it turns out my reservations were well-founded.

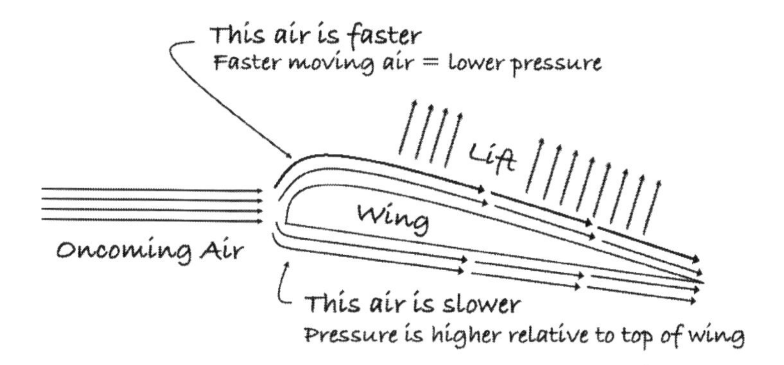

While digging into this question, I found a fascinating article from February 2020 in *Scientific American* written by Ed Regis. The title of the piece is what caught my eye: "No One Can Explain Why Planes Stay in the Air" It's a lengthy but informative story of how much debate still exists about why an airplane *really* flies. One thing Regis points out is that this idea—that air parcels on the top of the wing need to bust a move to reconnect with their long-lost parcel of friends who took the shorter route along the bottom of the wing—is not supported by the evidence. He notes that there is "no physical reason that the two parcels must reach the trailing edge simultaneously." In other words, there isn't some mysterious bond that draws them back together after they've been disrupted by the passing wing. And in any case, he further explains, experimentation shows that the air on top of the wing moves faster than this theory of lift can account for. That fact notwithstanding, Bernoulli's principle *does* hold—the faster air over the top of the wing creates pressure lower than that under the wing, generating lift.

One of the other, and to my mind, simpler explanations of lift goes back to our favorite physicist from high school: Sir Isaac Newton. It's an explanation that is the go-to for the author of a book popular in the student pilot community: *Stick and Rudder*. Wolfgang Langewiesche (pronounced long-gah-VEE-shuh) wrote the book in 1944, and while the style of writing and language in the book give away its age, it still is an insightful and helpful read. Langewiesche explains on page nine of his 384-page book the following: "The main fact of all heavier-than-air flight is this: *the wing keeps the airplane up by pushing the air down*" (emphasis his). This explanation is why I alluded to Newton at the beginning of this paragraph: Langewiesche explains lift as a product of Newton's third law: that for every action, there is an equal and opposite reaction.

If the wing pushes air down, the air must push the wing up.

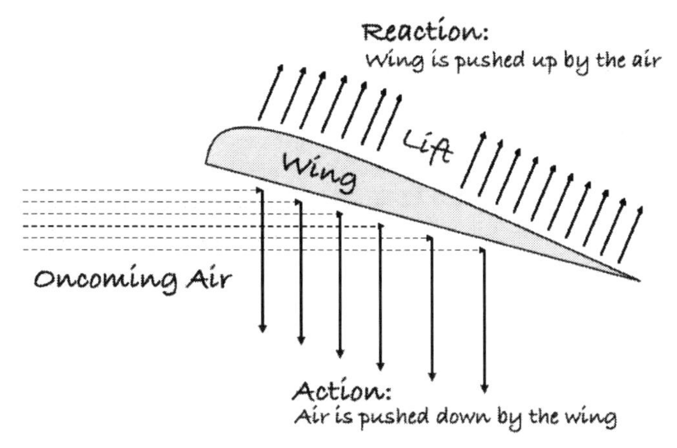

The most common demonstration of this explanation takes us back to our youth when we first put our flat hand out the window of a moving car and found that the greater we tilted the leading edge of our hand up, exposing more air to our palm, the harder our hand and arm wanted to fly upward.

So, what's the reason for lift?

Regis, by the end of his *Scientific American* article, and the FAA authors in the *PHAK*, thread the needle between these two theories of lift and explain that evidence suggests lift is created both because of Bernoulli's principle *and* because of Newton's third law. In fact, Regis notes, there seems to be some sort of synergy that exists between these forces and their effects. A sentence near the end of his article reminded me of Clarke's third law regarding technology and magic: "There seems to be a hint of magic in this synergy," Regis says. Indeed.

Where does all of this leave us when it comes to understanding why a wing creates lift? Here's what I'd say:

- A wing creates lift because air flow over the top of it is faster than the bottom, creating a difference in pressure between the top and bottom of the wing that draws the wing upward.
- A wing creates lift because when angled up, exposing the bottom of the wing to the oncoming air, it pushes the air down and the air pushes the wing up, driving the aircraft upward.

But there is an important condition that must be met regardless of which explanation speaks to you best: lift is only produced if the airflow over the top of the wing is smooth or mostly smooth. The more turbulent the air on top of the wing gets, the less lift and more drag it produces, until it doesn't produce any lift at all. This point is critical to understanding what sent us down this very interesting rabbit hole to begin with: stalls.

Chapter 6
Why Planes Stop Flying - Stalls

Understanding a stall begins with understanding something called angle of attack (AOA). As you might expect, AOA is, well, an angle. And an angle is created by the intersection of two lines. So, what two lines define an AOA? Let's go back to our earlier discussion where we defined the wing chord. This is one of our lines. Now imagine drawing a line that represents the direction from which air is hitting the wing, something we will call the relative wind. This is our second line. The intersection of these two lines defines our angle, and that angle is called the angle of attack.[19]

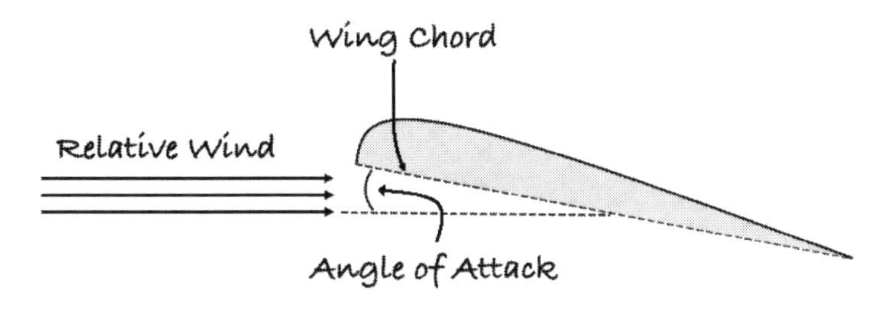

[19] I tried to find out why this angle is called the "angle of attack." Why the use of the word "attack"? I could find no definitive answer on this. It almost certainly was used in some form in the nautical world before the

The relative wind is the wind the wing *perceives* as it moves through the air. I say "perceives" because a relative wind can exist on a completely calm day. The airplane could be gliding with no engine and through air with no wind and yet the movement of the plane through the atmosphere will create a *relative* wind (much like, on a calm day, "wind" flows around your car when you drive it). In almost all cases, virtually all the relative wind in most aircraft (gliders being an exception) and in most situations is due to the engine—by pulling the aircraft through the air, the engine creates a wind over the wing. In fact, you may never have considered it, but an engine on an aircraft exists for one reason: to provide a relative wind to the wing. No relative wind, no lift.

Perhaps a clearer example of relative wind comes not from fixed-wing aircraft, but from rotary-wing aircraft. The blades on top of a helicopter are nothing but individual wings. The engine turns these "wings" to get them moving through the air, thus creating a relative wind. By moving the "wings" of a helicopter without moving the fuselage, lift is still created without any movement by the body of the helicopter, enabling vertical takeoff.

As you might expect, the AOA, which describes how the relative wind is hitting the wing, is an important part of flying—and controlling—an airplane. If the AOA is in an acceptable range, the wing will produce lift. But if it gets too large, air will not flow smoothly over the top of the wing; it will become turbulent and "separate" from the surface of the wing. When this

---

aeronautical one since a sail has an angle of attack relative to the wind. But the way I've always chosen to interpret it is that it's the angle at which the wing "attacks" the air.

happens, lift is lost, and the wing is stalled. This "too large" angle of attack is called the *critical* angle of attack and, in most aircraft, is between 15° and 20°. You've probably seen a representation of the physics of a stall if you've ever watched swiftly moving water passing over rocks in a stream. Water—a fluid like air—flows smoothly over the gently curved rocks (and, if you could measure the pressure over those rocks, it would be lower than the pressure in water passing unperturbed nearby), but the more surface area of a rock that is exposed to the oncoming fluid, the more roiled the water behind it will become. This is important in the context of our wing and the air moving over it because turbulent air doesn't create the reduced air pressure that smooth air does. And that reduced air pressure is one of the factors that creates lift.

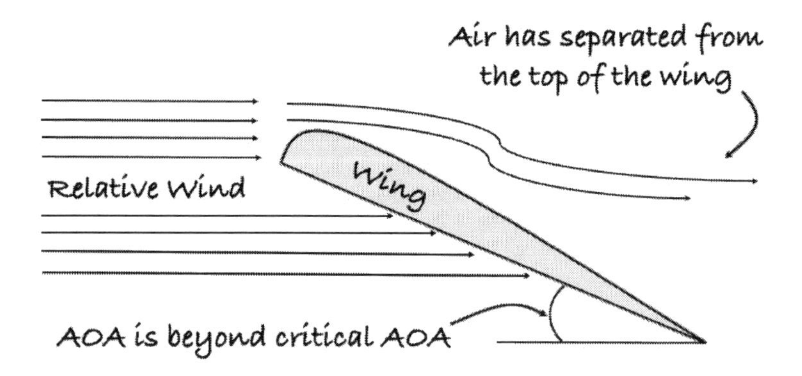

Our friend Wolfgang of *Stick and Rudder* fame gives a bit of a different explanation of why a stall occurs, and it's one that is consistent with his Newton's third law approach to lift. He notes that "whenever a wing meets the air at too large an Angle of Attack, and tries to wash it down too sharply, the air fails to take the downward curve." This, in turn, leads to the separation of the air from the top of the wing; it simply "burbles and breaks away from the guidance of the wing's top surface." It's this smooth bending of the air beneath the wing and smooth air over the top of the wing that yields our equal and opposite reaction, driving the wing upward. Without that smoothness, we don't have lift. So, as you can see, both versions of stall explanations end up in the same place: when the critical angle of attack is exceeded, air above the wing becomes turbulent, and the wing stops flying.

And what's amazing—and this can take you a bit to get your head around—is that a stall can occur at any attitude (i.e., the plane can be in any position relative to the ground) and at any airspeed. This point is emphasized in your training because even though a stall can occur at any airspeed, we often talk about the onset of a stall in terms of a particular airspeed.

For example, in a Piper Cherokee Warrior III Pilot's Operating Handbook (POH), the stall speed of a fully loaded aircraft is listed as 50 knots with no flaps.[20] Once you know this and you know the dangers of stalling low to the ground, you will remind yourself to stay above this speed. Not doing so could be deadly. But that implies that if you are above 50 knots no matter what, you're always safe. Unfortunately, that's not true. Remember, a stall can occur at *any* airspeed and in *any* attitude. The 50 knots only applies to an aircraft in certain conditions—namely level, unaccelerated flight, at a particular weight, and with specific atmospheric conditions.

---

[20] A POH is essentially an instruction manual for an airplane. It's worth reading cover to cover the one for the airplane you are flying. POHs are generally easy reads and help you get to know all you need to if you are to safely fly the airplane.

If an airplane can stall at any speed, why do we have a stall *speed* (rather than stall AOA) listed in our POH? Because in *most* flight situations, airspeed is a good proxy for angle of attack, and since all aircraft have airspeed indicators and most do not have AOA indicators, it's the best we can do. A booklet by AOPA called *Keep the Wing Flying*, highlights the question of airspeed as a proxy for AOA by noting that "airspeed can serve as a surrogate for AOA near the middle of the flight envelope, but the approximation becomes progressively worse at higher bank angles and/or more extreme pitch attitudes." In fact, the use of airspeed as a proxy for critical angle of attack is so well-accepted that nowhere in my POH is the term "critical angle of attack" used. More interestingly, and this is a bit hard to believe, "angle of attack" isn't even used.

Airspeed is only a proxy

Airspeed, then, becomes our go-to metric for safe flight. But, getting back to the "stalls can occur in any attitude and at any airspeed" mantra, don't fall into the fallacy of believing that just watching your airspeed will protect you from a stall.

Imagine flying a surprisingly rugged aircraft at 120 knots, straight and level. The stall speed for the aircraft is 50 knots. The shape of the airplane's wing allows it to stay aloft when the relative wind is hitting the wing at a very low AOA—let's say 3°. Suddenly, a bird flies in front of you. Without thinking, you jerk back on the yoke. If your power setting doesn't change almost instantly (and, depending on the engine's power, even if it did), the yank back on the yoke will make the angle between the relative wind and the wing chord excessively high. The critical AOA has been exceeded. All smooth air over the top of the wing is gone; it's completely turbulent. Lift is no longer being produced. At the moment your aircraft stalls, your airspeed may still be reading well above 50 knots. This is what's called an accelerated stall, for what I think are obvious reasons. Aerobatic performer Mike Goulian has said that when he puts his high-performance propeller-driven aircraft into a stall to enter a snap roll, his airspeed is around 140 knots, which is much higher than the published "power off" stall speed of the plane.

You may say, "Wait a minute, I've seen an F-16 at an airshow come down the runway at low level and then pull straight up abruptly and it didn't stall. I've also seen it do very abrupt and tight turns and it didn't stall then either." Fair enough. But even that airplane would stall if it didn't have enough power during the maneuver to push the airplane through the atmosphere so that the angle between the wing cord and the relative wind was less than the angle of attack. The pilot of the F-16 has around 32,000 pounds of thrust at their command, so they can use that power to keep the wing moving through the air at an appropriate AOA. This is hard for people to envision because we want to

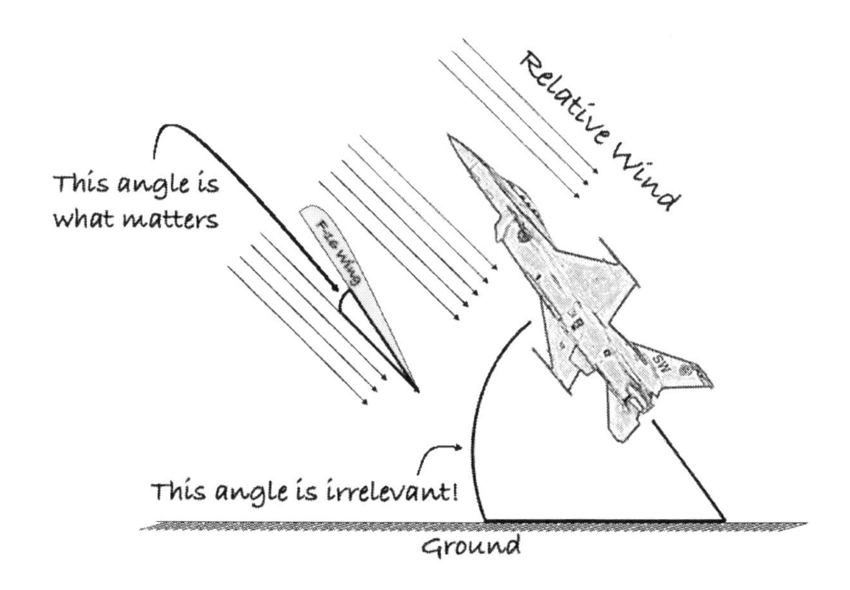

think of the angle of attack as having something to do with the airplane wing's angle relative to the ground. But that angle is irrelevant. When that F-16 pulls up, the wing may be 75° "up" relative to the horizon, but with enough power, the pilot can push the airplane through the air so that the wing "feels" the air coming from nearly head-on. The AOA stays in an acceptable range and the airplane keeps flying.

The Spin Cycle

While a stall is dangerous by itself, it becomes doubly dangerous if it develops into a spin. In fact, when you read reports of fatal accidents, you'll often see the phrase "stall spin" preceding "accident." This is *really* what I was referring to when I started this part of the book. Stalls can be dangerous, but stall-spins are much more so. This is because a stall that doesn't evolve into a spin in most light aircraft (and certainly all general aviation trainers) is generally benign and often recoverable. You'll see this when you do your first stalls early in your training. You raise the angle of attack—which usually means just raising the pitch from a straight-and-level attitude—and the plane will slow until you feel a shudder in the frame and the nose drops lazily. If the maneuver is coordinated, the amount of altitude lost is in the hundreds, not thousands, of feet, and the airplane simply recovers straight ahead, without dropping off to the left or the right.

You have probably already heard of "coordinated" maneuvers if you've taken any lessons. I'm sure you've been told early and often to keep the "ball" in the turn coordinator instrument centered, and if it's not, to "step on the ball" (step on the rudder pedal on the side the ball has drifted to) to get it centered. But that doesn't do a good job of describing exactly what's happening in coordinated flight. The *Airplane Flying Handbook* helps us out by explaining coordinated flight simply as a state where the "airplane's nose is yawed directly into the relative wind."

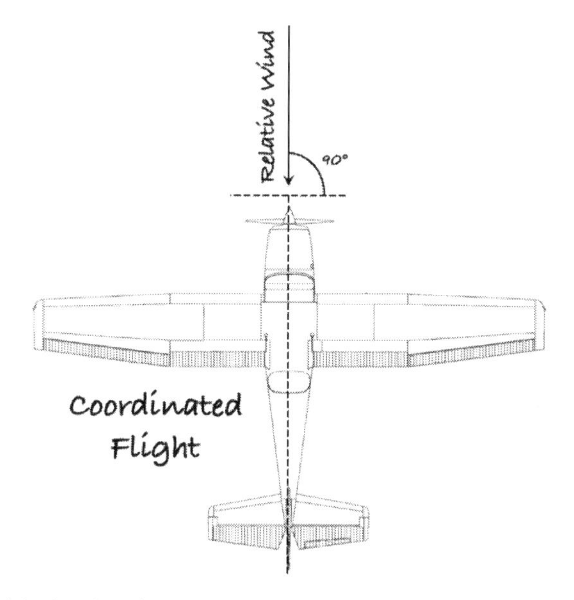

If this is the definition of coordinated flight, then you can probably surmise what uncoordinated flight is—a state where the nose is *not* yawed directly into the relative wind. You can be uncoordinated in both level flight and in a turn.

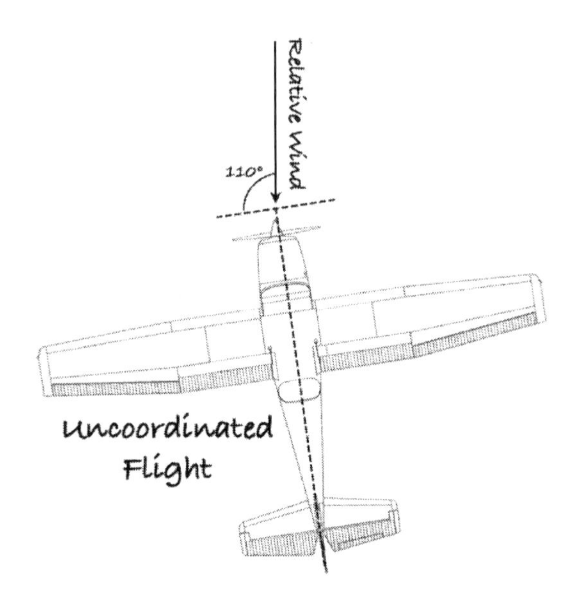

If you are uncoordinated in a turn, it means that the tail will either be flying to the outside of the turn or falling to the inside. Said another way, it means that, like the nose, the tail is not pointed into the relative wind. If the tail is on the outside of the turn, the airplane is in a skid. This should make sense as it's much like the skidding that might occur when a car is in a sharp turn at high speed. In this scenario, the vehicle "fishtails," meaning the rear wheels swing to the outside of the intended path, just as the tail of an airplane does in a skid (although for a different reason!). If the tail is to the inside of the turn, then you are in a slip; the tail has "slipped" toward the center of the circle the turn would draw if you continued around for 360°.

A turn of any meaningful bank will always require rudder if it's to stay coordinated. But why? It's because of something called *adverse yaw*. The word "adverse" is defined as something unhelpful or harmful, and its use in this context is completely appropriate. When you rotate the yoke in an airplane to initiate a turn, the aileron on the side in the direction you wish to turn (i.e., the left aileron when doing a left turn) goes up and the one on the opposite side goes down.[21] In a left turn, the up aileron reduces the lift on the left wing while the down aileron increases lift on the right wing. We'll discuss later why a down aileron causes the creation of more lift, but for now I'm hoping it's obvious that more lift *must* exist since this wing is rising to initiate the turn.

With that in mind, remember when I said increasing lift also increases drag? Well, so it is in this case. The right wing (again, in a scenario where you are making a left turn) now has more lift, but it also has more drag. This drag wants to "hold back" the right wing, which causes the airplane to want to yaw to the right. This is our adverse yaw. It's adverse because it's trying to pull the nose of the plane in the opposite direction you want to take it. That's why we need to help the plane along by applying rudder in the same direction as the turn.

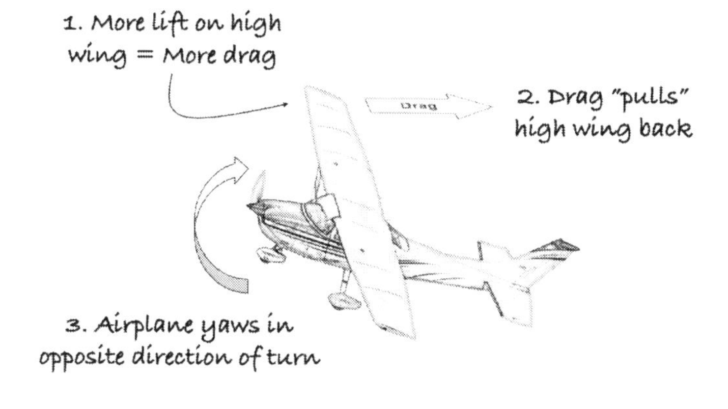

[21] For those very new to aviation … the ailerons are the movable surfaces on the end of the wing.

Keeping an airplane coordinated in a turn by applying rudder appropriately is an important part of being a safe pilot, especially at low speeds and high angles of attack. This begs the question of what happens when your airplane *isn't* coordinated when it's at *very* slow speeds and high angles of attack—when it's entering a stall. To answer that question, let's imagine entering a stall in an uncoordinated manner with the airplane yawing to the left; the tail is carving a path in the sky to the right of the nose if you are viewing the airplane from behind. When the stall happens, the nose will drop to the left (the direction of the yaw). The right wing will rotate upward. At this point, both wings are stalled (remember, that's how we got into this mess in the first place!), but the right wing is less so than the left.

As we just discussed when diving into adverse yaw, the more lift a wing generates, the more drag it generates too. Another way to say the same thing: the higher a wing's angle of attack, the more drag it produces. In our scenario where a spin develops, both wings are stalled, but the low (left) wing is *really* stalled. It has a higher angle of attack than the right wing and is therefore generating more drag and less lift.

You may be wondering though: Why does the low wing have a higher angle of attack than the high wing? It's because the high wing is moving through the air faster than the low wing because it is having to "race" around the outside of the turn the airplane is in. Therefore, the relative wind is hitting the wing at a shallower angle than it's hitting the low wing, which is moving through the air more slowly. In other words, while it has exceeded the critical angle of attack, the right wing hasn't exceeded it as much as the left wing. Also, the right wing's lower angle of attack means that it has *less* drag than the left wing as it speeds through the air in an arc. And while the right wing isn't producing enough lift to sustain flight, it has more than the left wing which causes it to keep rising. Eventually, the airplane will "nose over" to the left as the right wing continues to come up. As the airplane rotates around the dropped wing, this very unfortunate differential in the level of stall in each wing puts the airplane into a downward corkscrew path. You have entered a spin.

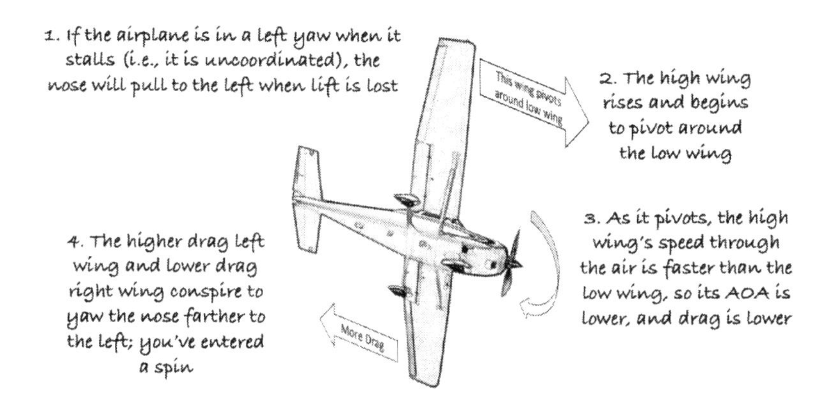

Spins in many aircraft are recoverable. But you need a good amount of altitude. This is why a stall-spin close to the ground is so dangerous and why it's *so* important to keep an airplane coordinated when slow in a landing configuration or in a high-power setting and climbing out on takeoff. Watch that ball!

Why all this time on stalls?

I've spent some time getting into stalls and spins and what causes them. But that's because stalls result in a disproportionate share of fatalities. According to AOPA in the *Keep the Wings Flying* publication I mentioned earlier, while stalls were only a factor in 10% of accidents between 2000 and 2014, those accidents accounted for nearly 24% of fatalities. Looking at it another way, AOPA notes that half of all stall accidents resulted in fatalities.

The good news is that the trend in fatal stall accidents is favorable. They are happening less frequently because of the FAA's work to ensure pilots are well trained to recognize when a plane is entering a stall and how to recover from one. There is still a long way to go, though. I find it telling that in AOPA data 20% of general aviation fatal fixed-wing accidents in 2014 were due to stalls (the peak over the 14-year period was around 29%). However, only 0.5% of commercial fatal fixed wing accidents were due to stalls.[22] This is both disheartening and encouraging at the same time. It's disheartening because of the losses that we see in general aviation and the difficulty in moving the needle on the fatality metric. It's heartening because the low commercial number indicates that training can help a *lot.* Yes, many aircraft used for commercial purposes do have some cockpit technologies that help to prevent them from entering stalls. And, as I noted earlier, they often have two crew members; two brains engaged in the flight are better than one. But the physics that apply to your Cessna 172 also apply to a 787 Dreamliner. Well-trained crews make a HUGE difference.

In your training, spend a lot of time getting a feel for what slow flight feels like and what too-slow flight (at the edge of a stall) feels like. And then commit yourself to recurrent training just like the big boys and girls at the airlines do. The FAA does mandate private pilots do recurrent training, but it's only required every 24 months. That's a long time to go without a refresher. And it can hardly be called a refresher; the FAA notes the flight review requirement is met if you engage in just one hour of

[22] Note here that "commercial" doesn't refer just to scheduled airliners that carry paying passengers from point A to point B. It also includes chartered flights, freight hauling, or any other flight where a pilot is paid for their service.

ground instruction and one hour of flight instruction. That's not very much at all. It can lead to a "check the box" mentality among many (dare I say most?) pilots. Think of it this way: Would you get on an airliner if you knew that the pilot had spent 55 hours of time (about the average time it takes a student pilot to become a certificated pilot) to get his ticket to fly passengers and then only spent two hours with a local instructor every two years to brush up on things? If so, then you are braver than me.

I'm not advocating that the FAA overregulate here; the U.S. general aviation system is unbelievably free of restrictions, making it incredibly pilot-friendly. But make it a point to take the FAA's requirements and add your own safety factors. I've committed to hiring an instructor every six months to try to keep myself sharp. And because the FAA doesn't prescribe what you must cover with an instructor during your flight review, give some thought to asking your instructor to challenge you on the sorts of things that could hurt or kill you or your passengers. Certainly, stalls should be high on this list.

Chapter 7
A Weighty Subject: Weight and Balance

It's probably intuitive to you that the weight of an aircraft matters. After all, it's one-fourth of Cayley's four forces acting on an aircraft in flight. But it matters in ways most who start pilot training never consider.

First, let's talk about the easy part. Yes, the heavier an airplane is, the "harder" it is to get in the air. Section 5 of any aircraft POH has tables or graphs that show takeoff roll at various weights, pressure altitudes (which we'll discuss in a later chapter), and temperatures. If we look at the takeoff distance table in the POH for a 1978 Cessna 172 we can see how much weight affects performance. At a weight of 2,300 pounds with pressure altitude at sea level and temperature at 20° C, the takeoff roll is 835 feet. At the same pressure altitude and temperature, but at a weight of 1,900 pounds, the takeoff roll is 540 feet, or 35% shorter. This is because it takes longer to accelerate the aircraft to a speed at which enough air is flowing over the wings to create enough lift to carry you and your aircraft into the wild blue yonder. And, by the way, the amount of air you need flowing over the wings to get airborne changes with weight too. That same table from the 172 POH shows that liftoff speed at 2,300 pounds is 52 KIAS (knots indicated airspeed). At 1,900 pounds, it's 47 KIAS. The faster air over the wing equates to more lift, which you'll need as the aircraft gets heavier. And a little airspeed goes a long way toward increasing lift. The lifting force of the wing is proportional to the square of an airplane's velocity, assuming other variables like angle of attack are held constant. For example, the lift generated from the wing of an airplane going 200 knots is four times greater than the lift of that same wing going 100 knots. Applying this fact to our earlier example means that the 52 knots necessary to get our heavier Cessna 172 off the ground equates to about 22% more lift than that needed to get our lighter 172 off the runway.

Flying with an overloaded airplane is something that is easy to get lulled into. I think a huge contributor to this is the number of seats in single-engine aircraft. Most have four. But many aircraft can't carry four full-sized adults if they carry any meaningful amount of fuel—and depending on how big your passengers are, it may not be possible to get airborne with a gallon in the tanks! The "invisible" nature of fuel is another thing that can contribute to taking off overloaded. Unlike baggage or a friend, it's not visible while you're preparing for a flight. Sure, you can look in the tanks to check the fuel state. You can also check the fuel gauges. But neither option typically gives you a very accurate fix on the amount of fuel in the tanks (unless they are completely full).[23]

---

[23] Analog fuel gauges don't provide enough precision for you to determine the weight in the tanks down to the pound. But digital cockpit instruments—like the Garmin G1000—will include engine and fuel readings. When used correctly, the fuel levels shown in such instruments are more accurate than "traditional" gauges because they calculate fuel left in the tanks by knowing the beginning state of the fuel and then subtracting fuel used by measuring fuel flow to the engine.

Another thing that lulls pilots into thinking they can fly heavy is that, in many cases, the plane will get airborne when overweight. If enough runway is available, giving the engine time to accelerate the aircraft to the point at which the wings can generate enough lift will, in many cases, allow the airplane to "limp" into the air; it will climb very slowly. If a plane does get airborne in a scenario like this, turns become especially dangerous. This is because a turn takes some of the lift pulling you up and "wastes" it by taking some of that force and pointing it in the direction of the turn. If the remaining vertical lifting force is less than the weight of the plane, you are in trouble.

This is, in fact, what led to the crash of a Cessna 172 in 2015 near Cleveland, Ohio. The plane had four people on board, all college students. Investigators determined after the crash that the airplane was overweight by at least 90 pounds, and probably more. Witnesses who were with the four young men before they boarded the plane said the pilot asked each person their weight and then did some "calculations in his head" before determining the flight would be safe to undertake. Once airborne, the NTSB report notes that the pilot "radioed that they were not climbing fast and that they wanted to immediately make a left turn to turn around." The report then explains what happened next: "The controller approved the left turn … the airplane began … [the] turn when it descended to the ground." This left turn was undoubtedly a stall turning into a spin. They needed every component of the lift being generated to be perpendicular to the wing—i.e., straight "up" while the airplane's wings were parallel to the ground. The minute the pilot turned left, he gave up some of that "straight up" lift and the plane could no longer fly. Don't "do the math in your head." Write the numbers down and add them up. Account for fuel and cargo. Your life depends on it.

V_a – The most mysterious airspeed

Weight also matters when it comes to guarding against damaging stresses on a plane when it's in flight. Consider an airplane in cruise flight. A heavier aircraft will fly at a higher AOA than a lighter aircraft under the same conditions (e.g., at the same airspeed). This is because more lift is needed to keep a heavier airplane flying, and increasing the angle of attack increases lift. This is a fundamental concept and is important to know for many reasons. One of those reasons is related to an important speed listed in your POH: V_a, or maneuvering speed. This is the speed above which abrupt, full deflection of a single control surface might result in structural damage to the aircraft.

Now that you know what V_a is you may have noticed that the V_a numbers for different weights in your POH seem counterintuitive. For example, in a Cessna 172S POH, the V_a for a plane weighting 2,550 pounds is listed as 105 KIAS. At 600 pounds lighter, V_a is 90 KIAS.[24] *But wait a minute*, you may think, *wouldn't flying faster when the plane is heavier put more stress on the airframe?* Good question.

This is where the higher angle of attack of a heavier aircraft comes into play. A higher AOA means that the wing on a heavy airplane in straight and level flight is closer to a stall. For example, let's assume the critical AOA for an aircraft is 18°. Let's further assume that a heavy version of this aircraft flies in cruise at 7° AOA at a given airspeed, and a lighter version flies at 3° at the same airspeed. That means that the heavy airplane is 11° from a stall while the lighter airplane is 15° from it. Why is this

---

[24] The fact that maneuvering speed changes with weight is why V_a is not marked on your airspeed indicator; it's not a constant figure.

important? Because there's a correlation between lift, AOA, and G-forces, and that correlation dictates the stresses on your airplane when an abrupt change in AOA happens.

Imagine you're piloting your heavy airplane in straight and level flight at 100 knots. Your AOA is 7°; you and your plane are experiencing one times the force of gravity (1 G) as lift matches weight. Now, imagine that you double your AOA to 14° because you abruptly pull back on the yoke. In this scenario, your lift and

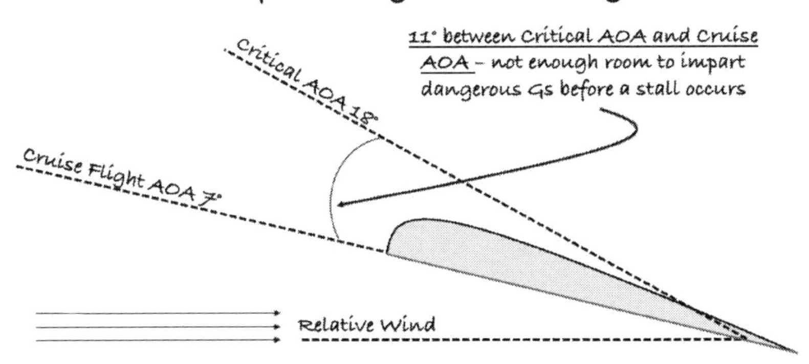

G-forces will also double. You'll feel 2 Gs, but it's not dangerous to you or your airplane. This is because most training aircraft are certified to withstand +3.8 Gs.[25] Now, imagine your angle of attack triples to 21°. Will your Gs also triple? Well, they would ... except at 21° the airplane is past its critical AOA. In other words, it's in a stall. And a stall means that the wing is no longer producing lift; there's no load on it at all. No lift, no Gs. No Gs, no worry about breaking the airplane. Essentially, the higher

---

[25] In fact, part of the specification of a "normal category" aircraft is that it has design load limits of +3.8 Gs. A utility category aircraft is designed to withstand +4.4 Gs and an acrobatic category aircraft is designed to withstand +6 Gs.

AOA needed to maintain level flight in a heavier aircraft doesn't leave enough room for the AOA—and Gs—to increase to a point that the structural integrity of the aircraft is compromised.

If you run the same thought experiment with our lighter airplane, you'll end up in a different place. Recall that we said that because it is lighter, if it is also flying at 100 knots, it can cruise at a lower AOA of 3°. If that AOA doubles to 6° because of turbulence or a control input, then the plane experiences 2 Gs. No problem. Now, assume it triples from 3° to 9°. Lift triples too and G-forces are now at 3 Gs. Again, no problem. Finally, assume AOA quadruples to 12°. The airplane now experiences 4 Gs. This *is* a problem as it exceeds the 3.8 G limit. But here's the important point: unlike the heavier airplane, the lighter plane *can* get to 4 Gs because at 12° the AOA is still well below critical. In other words, the plane *didn't* stall before it exceeded design structural limits.

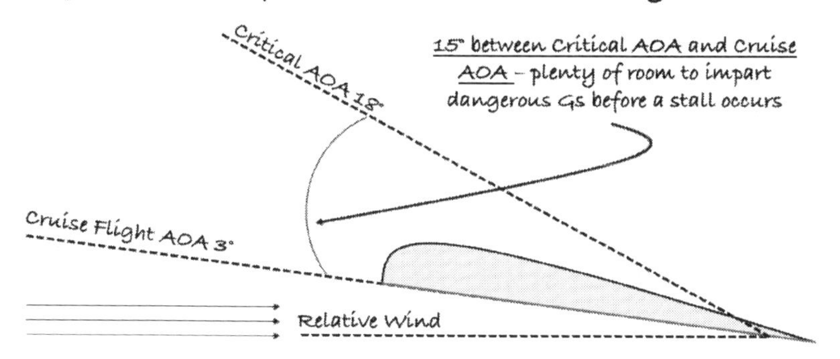

The V_a speed for a lighter aircraft is lower than for a heavier aircraft to force you to fly the airplane at a higher AOA. By flying a lighter airplane at or below V_a, you have mimicked what flying a heavier aircraft is like by forcing the wing into a higher AOA and providing that built-in stall protection if a sudden change in AOA occurs due to a control input.

***

While I was writing this book, there was an accident in Mexico involving a crop duster that likely shows in dramatic and tragic fashion what happens when V_a is exceeded.[26] As you probably know, a crop duster is an aircraft that has a hooper—a container—that can hold fertilizers and pesticides which can be sprayed or dropped from the plane while it's in flight. Crop dusters can be an efficient way to apply a treatment to a large swath of land.

In the case of this accident though, the plane wasn't carrying any payload for agricultural purposes. Rather, it carried what appeared (based on a video of the event) to be a dry, pink powder. The substance was pink because the purpose of the flight that day was to be the centerpiece of an announcement of the gender of a couple's baby at their gender-reveal party. While the expectant couple was standing outside in front of lit, block letters that spelled "Oh Baby," the crop duster approached from behind them and, as it neared the crowd, dropped the pink powder, revealing that a baby girl was on the way.

As the load was dropped, the pilot pulled up aggressively. Such a maneuver would have imparted significant G-forces to the airplane. At this moment, the video captures the left wing "fold up" from its connection point at the fuselage as the plane's nose points skyward. With no lift on one side of the aircraft, the plane twisted in an arc before impacting terrain, killing the pilot but, fortunately, sparing everybody on the ground.

---

[26] I say "likely" because an investigation into the accident had not been concluded when this book was published.

While we can't know for sure, it may have been possible for the pilot to do such an aggressive maneuver without damaging the airplane and killing himself if the load he was carrying was *not* dropped. The aircraft would have been meaningfully heavier, and V_a would therefore have been much higher. In other words, at the heavier weight, the plane could have maneuvered aggressively at a higher speed because it would have stalled before a harmful G-load was achieved.[27] But as the pink substance left the plane, the aircraft became quickly lighter, which would have reduced V_a in an unexpectedly rapid way. In fact, if my theory on the cause of the accident is correct, the speed the plane was at when the pilot pulled back quickly on the stick was beyond the "new" V_a in the lighter airplane. This means the pilot was able to exceed the design load limits of the airplane with his attempted climb; the G-forces became structurally damaging before the aircraft stalled.

While this story is harrowing, and understanding maneuvering speed is important, don't obsess about it. Remember, the speed is relevant when executing a "full or abrupt" control movement. Such movements, under normal flight conditions, should be rare. I can't think of a single time I've had to make such an input while flying.

Balance

As I mentioned earlier, it's intuitive that weight should be a factor in the performance of an airplane. Cars don't go as fast when they have a lot of weight in them, and we don't move as

---

[27] Note that in this scenario, a stall would probably have resulted in a similar outcome. It appears from the video that the plane would have been too low to recover.

fast when we have a lot of weight on us. Our experiences outside of aviation translate well to the flying world. But unlike our earthbound references, *where* the weight is in an airplane also matters. It matters a lot. This is the balance part of "weight and balance."

To understand why it matters, let's go back again to a little high school physics that at first may not seem related to our aircraft balance discussion but I promise is. You may remember learning how having a lever and a fulcrum (a point on which a lever is pivoted) makes lifting a heavy weight easier than if you tried to lift it without a lever. In fact, Greek mathematician and physicist Archimedes, who did much of the early work to understand why levers do what they do, is believed to have said some variation of "give me a lever long enough and a fulcrum on which to place it and I shall move the world." If Archimedes did indeed try to move the world with a lever, we'd describe the end with the world on it the "weight end" and the end on which he pushed the "effort end."

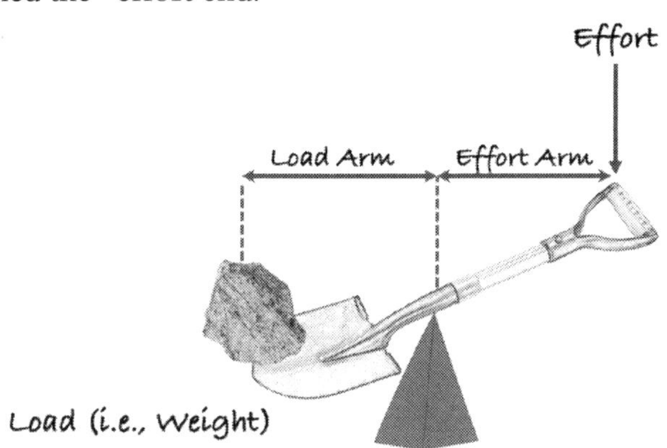

Even without studying physics, you have probably learned the lesson at some point in your life that having the fulcrum closer to the weight makes the effort easier. For example, if you've ever tried to use a shovel to get a rock out of the ground, you may have used a log as a fulcrum and placed it under the neck of the shovel (i.e., close to the blade of the shovel and the rock you are trying to lift) to make it easier to pry the buried rock from the Earth. If you had instead put the log (fulcrum) closer to the effort end, the work would have been harder.

The distance between the weight or "load" to the fulcrum is, uncreatively, called the "load arm." The distance between the effort and the fulcrum is, also uncreatively, called the "effort arm." The longer the effort arm and the shorter the load arm, the less effort you'll need to lift your rock.

This relationship between the location of a fulcrum and the

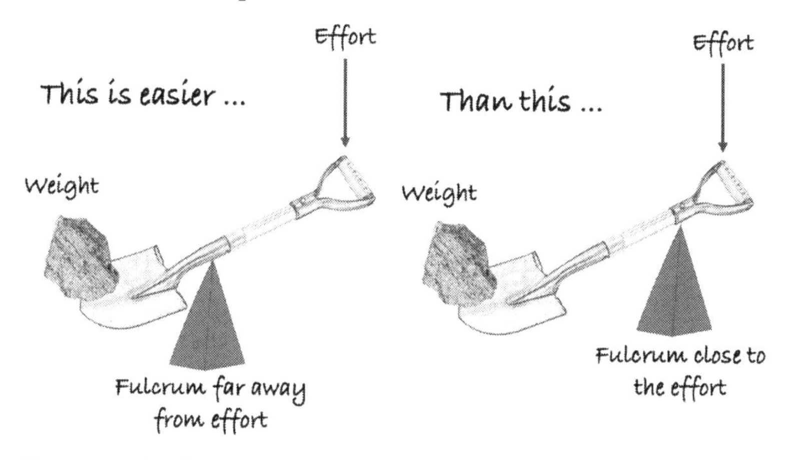

effort required to move something brings us back to our aircraft—because the fulcrum in our example here is related to the center of gravity (CG) of an airplane. In the *PHAK*, the FAA defines the CG as "the point about which an aircraft would balance if it were possible to suspend it at that point. It is the mass center of the aircraft or the theoretical point at which the entire weight of the aircraft is assumed to be concentrated." Said another way, if you placed an airplane on a fulcrum so that it was perfectly balanced—the nose and tail were level, the wings were level—the point on the aircraft where the fulcrum was placed would be the CG. The location of the CG in an airplane is measured as a distance from some reference plane called a datum plane. Often in a single-engine aircraft, the datum plane is the firewall—the wall between the engine and the rest of the airplane. For example, a Cessna 172 I've flown has a CG 41.7 inches from the datum when it has no passengers or fuel. This puts the CG on a line that runs roughly through the center of the front seats from one side of the aircraft to the other. If you placed a fulcrum under the airplane 41.7 inches back from the firewall and halfway between the left and right sides of the fuselage, the airplane should balance perfectly atop it.

Thinking of the CG as a fulcrum helps to illustrate why, in addition to understanding the total weight of an aircraft, understanding how that weight is distributed matters to its safe operation. This is because the CG is the pivot point about which

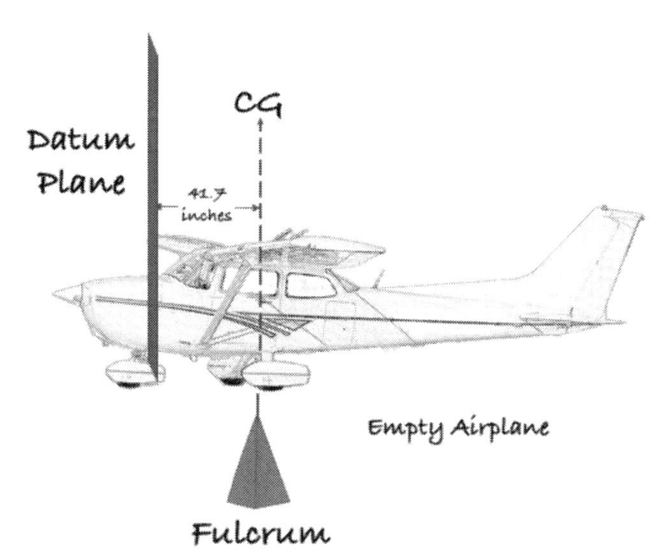

the airplane moves in pitch, yaw, and roll when you move the elevator, rudder, and ailerons. And when it comes to the rudder and elevator, they are exerting force about the CG with the fuselage acting as a lever. The tail is the "effort end" of that lever. The "weight end" is the nose.[28]

To explain further, let's go back to the 172 from the last paragraph and assume that with me flying in it alone with 50 gallons of fuel and no baggage the CG is at 42.1 inches. If I add two 165-pound adults to the back seats, what do you think will happen to the CG? If you guessed it will be farther back on the

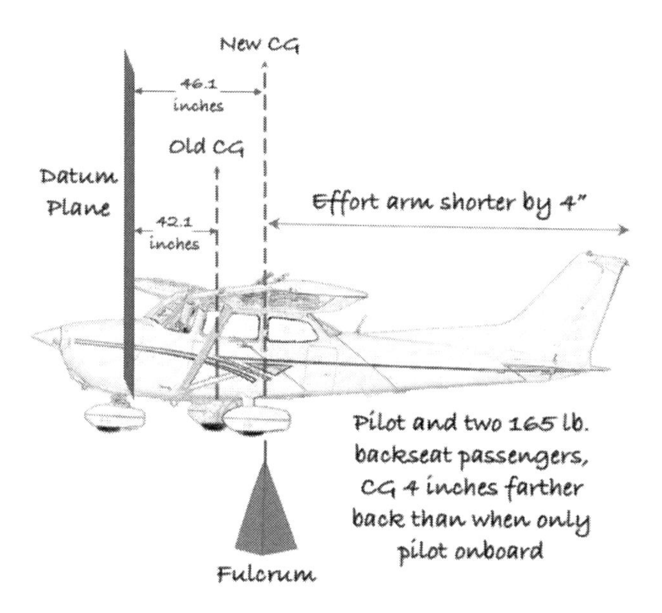

airplane, you're right. In fact, the CG in this scenario would be at about 46.1 inches; if we were to place our plane on a fulcrum, the point at which it would balance would be four inches farther back from the datum plane than when it was just me in the airplane.

---

[28] This is a simplified explanation of how CG and control are related. To get into the details would require more discussion of the location of the center of lift in relation to the center of gravity and something called the "tail down force." But the general principles described here are sufficient to understand how a center of gravity closer to the tail degrades control.

Knowing this, do you think that in the case where I have my two friends in the back seat the rudder and elevator will be more, or less, effective? If you said less effective, you're right. If I deflect the rudder all the way to the right at a given airspeed, a force is applied to the tail that swings the nose to the right. If the CG is farther back and I deflect the rudder all the way to the right at the same airspeed, the same force will be applied to the tail, but it won't swing the nose as effectively to the right. This is because we've moved the fulcrum closer to the "effort end" of our lever (the tail) and, just like a shovel with a fulcrum closer to the handle, the work is harder to do.

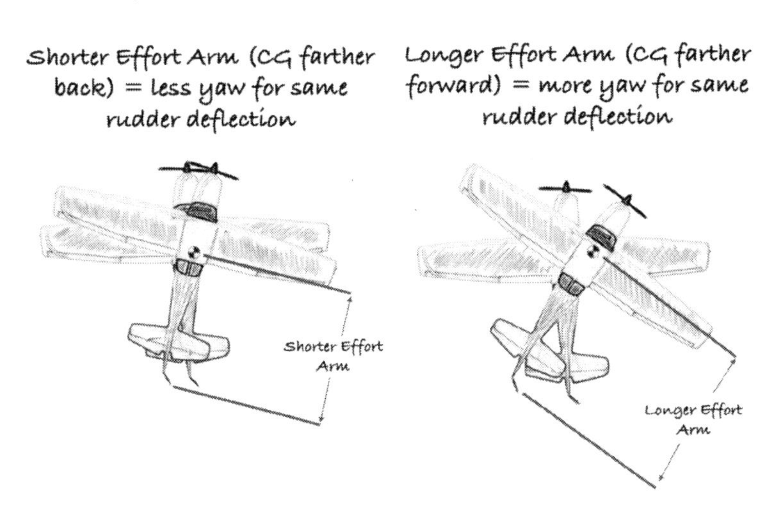

Speaking of our rock example, you may have seen a correlation between "load arm," which I described earlier, and the distance the CG is from the datum plane. Sure enough, the 46.1 inches from the datum to the CG in the plane with my two buddies in the back can be thought of this way, although it's not a perfect correlation since there's also weight in front of the datum plane: the engine. But since the engine doesn't change in weight from flight to flight and there's nowhere to put additional weight (like baggage) forward of the firewall in single-engine trainers, thinking of the distance from the datum plane to the CG in this way is just fine. This will make more sense, as will the terminology, when we discuss calculating weight and balance shortly. Importantly, though, the effort arm as described in relation to the CG and tail of the airplane *can* be thought of in its purest sense. In our example above, when two passengers in the back of the plane moved the CG back by four inches, the effort arm got four inches shorter.

In this example of a four-inch shift in CG and subsequent shortening of the effort arm, the reduced effectiveness of the rudder and elevator may not be that noticeable because the CG, even with two 165-pound backseat passengers and me in the front seat, is still within limits for the aircraft I referenced, although barely. But if the CG gets beyond about 47.2 inches, then control becomes more difficult. When I say that control becomes more difficult, it doesn't necessarily mean that the airplane can't get airborne and be maneuvered. If the aft CG limit hasn't been exceeded dramatically, the plane will likely fly. As with being overweight, this fact makes the dangers of flying an airplane with an aft CG out of limits even more insidious.

To illustrate that point, let's discuss the 2013 crash during takeoff of a de Havilland Canada DHC-3 "Otter" in Alaska. The aircraft got airborne, but the nose continued to pitch up until the aircraft stalled and then plunged, impacting the ground and killing all on board. The Otter is a tail-dragging airplane, but it's large as far as taildraggers go. The accident airplane had a turboprop engine that allowed it to carry heavy loads in and out of some of Alaska's shorter fields. On the day of the crash, it had nine passengers and a pilot on board.

As with all accidents, there were a few factors that contributed to the crash of this Otter. One was the improper use of flaps. Cell phone video taken out the window by one of the passengers and recovered from the crash site showed that the flaps were in a landing position (i.e., full down). This was not the appropriate setting based on the POH. That said, it's unlikely flaps set at "full" would have caused a crash for a pilot with the experience (over 7,000 hours) of the pilot flying that day. It's true that such a flap setting would have caused the nose to rise more rapidly than expected and would have made the aircraft fly slower, but it's probable the pilot would have realized this quickly and been able to use nose-down elevator to keep the airplane away from a stall until he was able to raise flaps to the appropriate setting.

But he couldn't do this because the center of gravity was aft of limits. In its investigation, the NTSB used some sophisticated analysis to estimate that the airplane "exceeded the maximum gross weight of 8,000 pounds by about 21 pounds." But more critically, the final report says the "CG would have been at least 5.5 inches aft of the 152.2-inch limit." By having the aircraft loaded in a way that moved the CG farther back, the pilot shortened the effort arm the elevator and rudder had to work with. He shortened it *so* much that even with full down elevator, he would have been unable to keep the nose from continuing to pitch up and ultimately lead the plane into a stall. That 5.5 inches may not seem like much. After all, the full length of a DHC-3 is 16 feet, five inches, or 197 inches. This means the distance the plane was out of its CG limit accounted for just 2.7% of the full length of the aircraft. But that's all it took to make the elevator completely ineffectual. In fact, the NTSB report notes that the "CG was so far aft of the limit that the airplane likely would have stalled even with the flaps in the correct position." A little bit of aft CG has a disproportionate impact on controllability. Don't mess with it.

One last point on this topic: I've focused on the dangers of an aft CG here because a CG out of limits to the rear of the aircraft is generally more problematic than one to the front. But that doesn't mean that a to-far-forward CG is benign. A CG beyond limits to the front of the aircraft can make it difficult to keep the nose up, and you're right back to control issues. This can be particularly problematic when landing. There are many accidents in the NTSB database that can be attributed to this type of out-of-limit CG.

Calculating weight and balance

Calculating weight and balance is straightforward. Part of the reason it's straightforward is that the only formula you need to know is this:

Weight (lbs.) x Arm (inches) = Moment (lbs.-inches)

With this formula and some tables and/or graphs in your POH, you can calculate the location of the CG in your airplane. Section 6 of your POH is where you'll find this information and that information is particular to your aircraft. While two Cessna 172s rolling off the assembly line might have the exact same weight and CG when empty, as aircraft are changed through the years, these measures change. New avionics, exhaust systems, the addition or subtraction of wheel pants[29]—all these things and more change the weight of the aircraft and its CG. Section 6 of the POH is the weight and balance fingerprint of your aircraft and should include any changes made to the airframe over the years to account for new weight and balance characteristics.

While the particulars of how to calculate the CG of your airplane for a given flight might vary depending on how the POH facilitates the calculation, fundamentally, you'll always be using our *Weight x Arm = Moment* formula. A Cessna 172R POH is typical. It shows a table with a sample aircraft to illustrate how a calculation is done. I'm going to use the sample aircraft numbers to explain the calculation.

ITEM DESCRIPTION	**SAMPLE AIRPLANE**	
	Weight (lbs.)	Moment (lbs.-ins./1000)
Basic empty weight	1639	64.4
Usable Fuel (at 6 lbs./gal.)		
53 Gallons Maximum		
35 Gallons (to reduced fuel tab indicator)	210	10.1
Pilot and Front Passenger	340	12.6
Rear Passengers	220	16.0
Baggage Area 1	48	3.4
Baggage Area 2		
RAMP WEIGHT AND MOMENT (add columns)	2457	106.5
Fuel allowance for engine start, taxi and runup	-7.0	-0.3
TAKEOFF WEIGHT AND MOMENT (Subtract fuel allowance from ramp weight)	**2450**	**106.2**

29 Wheel pants are the covers you'll often see on the wheels of aircraft. Wheel pants make a plane more aerodynamic, helping increase the speed and efficiency of an aircraft.

You'll notice that the columns in the Cessna POH include weight and moment. The arm part of *Weight x Arm = Moment* is missing. A few pages before this table, Cessna provides a diagram of the aircraft that shows the distance from the datum for each of the locations that appear under "Item Description." For example, the pilot and front passenger are 37 inches from the datum (at the midpoint of the seat position). So, in the table above, if you were completing it yourself, you'd multiply the weight of the pilot and front passenger by 37 to get the moment. And in fact, 37 in. x 340 lbs. = 12,580 in.-lbs. You don't see this exact number in the right-hand column as it's been rounded up to 12,600 in.-lbs. and (as noted in the header for the column) divided by 1,000. Dividing by 1,000 is just a way to make the numbers easier to manage.

The other way to get the moment is to use a chart provided in the pages of the POH. It shows weight on the y-axis and moment on the x-axis. There is a line for each position in the aircraft where weight is variable—i.e., pilot and front passenger, fuel, rear passengers and baggage areas. You find the weight for each position on the y-axis then follow a line to the right for that weight until it intersects with the position you are trying to find the moment for. Then you run a line down from that point to the x-axis. The reading there is the moment divided by 1,000. I find the charts in most POHs ridiculously hard to read. I therefore prefer to calculate the moment based on the actual position of the load as specified in a table or diagram in the POH whenever possible.

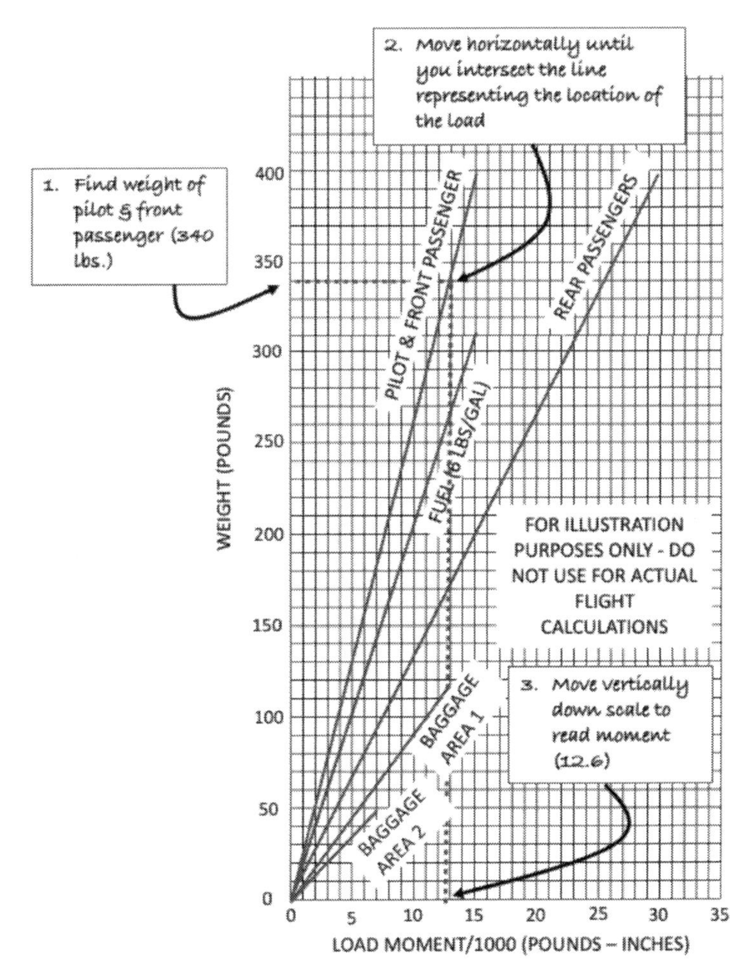

You follow this process to get the weight and moment for every load on the aircraft. You then add the weights and the moments listed in the columns and then use the grand total of each along with another chart to plot the CG of the aircraft and ensure it is within the appropriate envelope. In our example in the previous pages, the total weight is 2,450 pounds and the total moment is 106.2 lbs.-in. (x1,000). In the POH, there is yet another graph called "Center of Gravity Moment Envelope" that utilizes the two numbers you just came up with to determine if the airplane is within CG limits. You find the weight of the airplane on the vertical axis and then follow the line for that weight to the right until you intersect the line corresponding to a place on the horizontal axis representing the moment you calculated. If the point where the two lines intersect is within a bounded area on the graph, your aircraft is within CG limits.

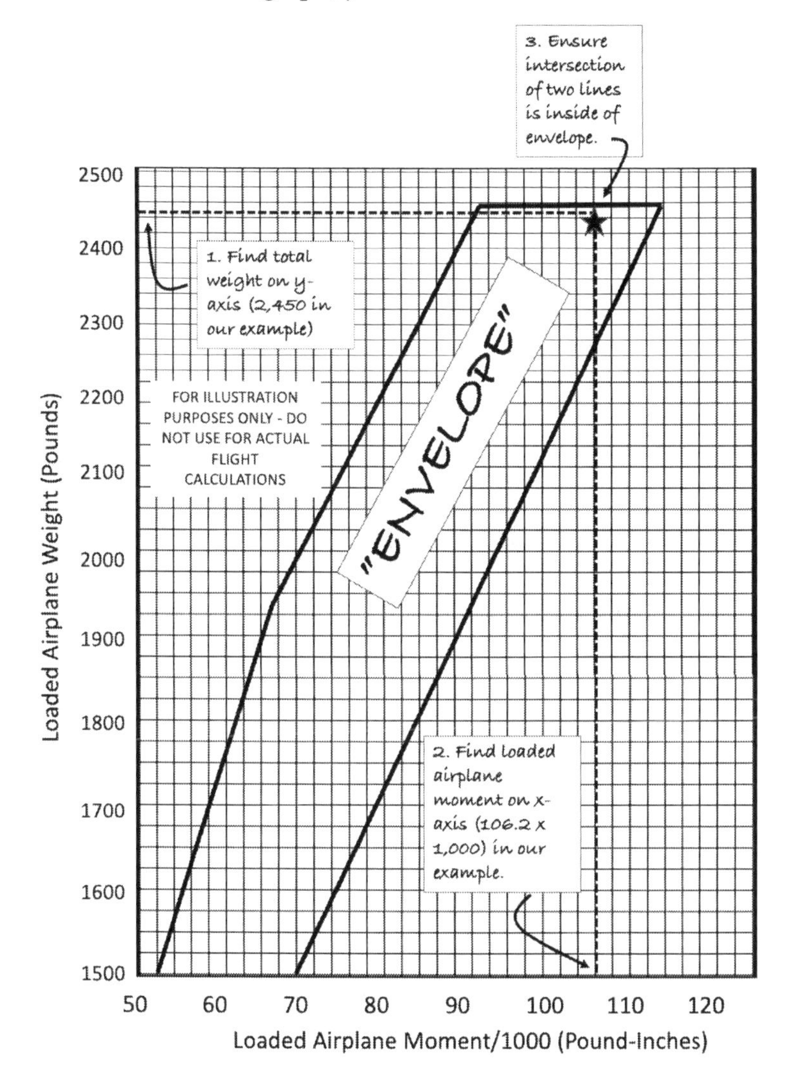

Note that when you calculate weight and balance, you'll often find the CG is right up against the edge of the boundary line on the inside of the envelope. Don't let this alarm you. This isn't uncommon. That said, *do* make sure the CG is *inside* the envelope.

I've described the math and process for determining whether your aircraft is within CG and weight limits. But I haven't really touched on the logic behind why the moment matters. To get to that, let's go back to our non-aviation example—a typical lever on a fulcrum with weights on either side at varying distances. If the lever is to be balanced, then the sum of the moments for each load on one side of the fulcrum must match the sum of the moments on the other side as illustrated below. In this example, I've multiplied each of the weights on the lever by their distance from the fulcrum to get their individual moments. You can see that on the left there is only one weight and therefore only one moment to worry about, and that moment is 50,000 lbs.-in. Using the same *Weight x Arm = Moment* formula, I then calculate the moments for the three weights on the right side of the fulcrum. When that's done and those moments are added together, I also get 50,000 lbs.-in; our lever is balanced on the fulcrum.

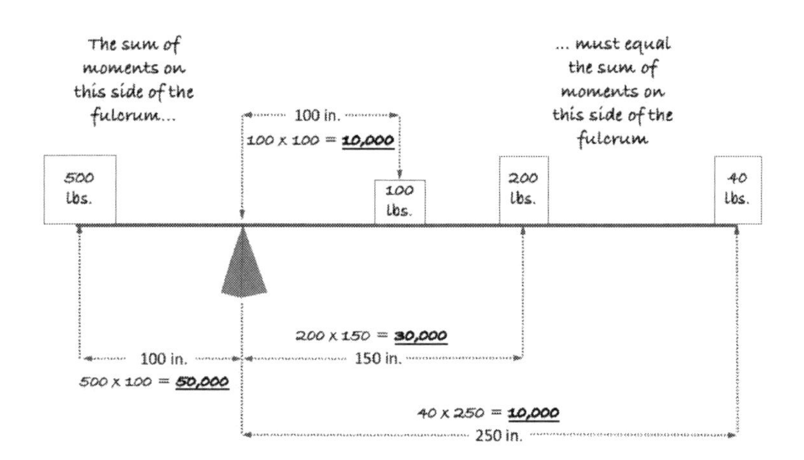

To understand what this has to do with an airplane's center of gravity, recall that the fulcrum in our example above represents the lever's center of gravity with the weights arranged as they are. Now, assume the weight on the far right is changed from 40 pounds to 80 pounds. Without changing anything else, our picture would now look something like this:

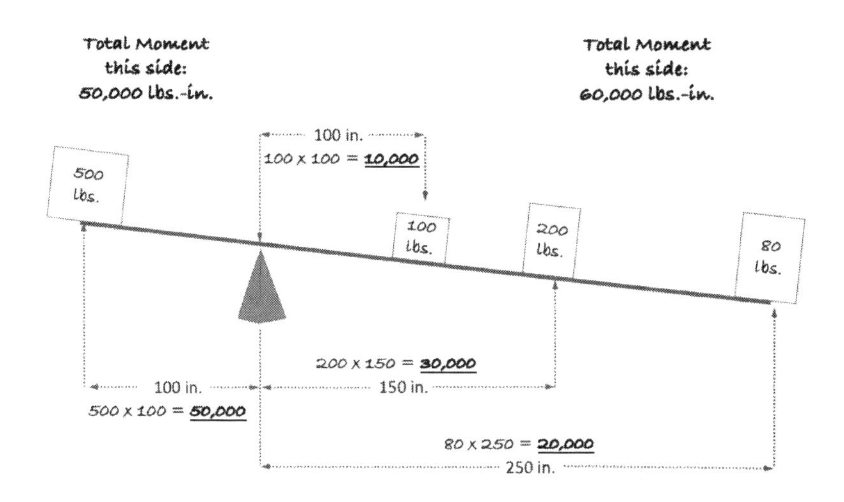

The total moment on the right is 60,000 lbs.-in. and on the left it's 50,000 lbs.-in. The lever is unbalanced. We could rebalance it by changing the weight on the far right back to 40 pounds or we could change the size and/or position of any of the other weights along the lever. But there's something else we could do … we could move the fulcrum. If we move the fulcrum to the right by 11.36 inches, we will be back in balance with the total moment on each side equaling 55,681.8 lbs.-in.

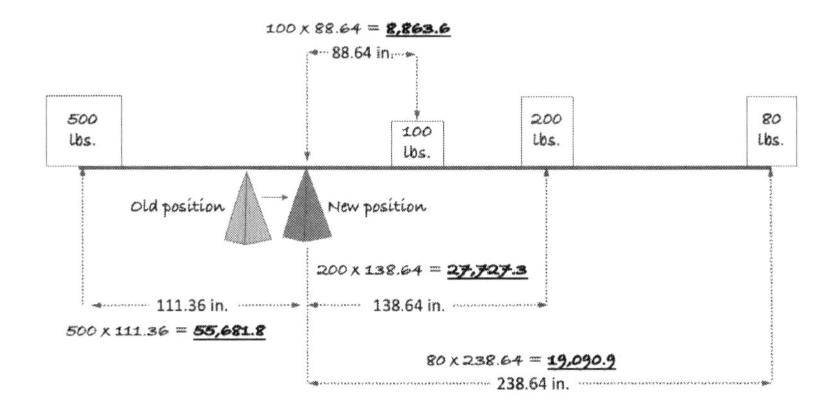

Our fulcrum represents the CG of the lever and weights if they are perfectly balanced. This is the key point to remember as we try to put the physics of this example into an aviation context. It would be great if, like a seesaw, we could just swap out one weight for another or change the location of a weight to keep the CG in a single location. If we could do that, we'd never have to worry about being out of CG limits. But this is impractical. Your fuel tanks can't be moved. You can't change the weight of Cousin Eddy. You can't move the engine or the location of the passenger seats. Each time we load the aircraft differently from one flight to the next, the CG will change. It will even change during the flight as fuel is consumed. When you calculate weight and balance using the *Weight x Arm = Moment* formula and the associated tables and charts in the POH, you are really calculating where you need to move the fulcrum to balance the aircraft—you are calculating where the CG will be. Knowing that the CG must move, aircraft are designed to allow some variability in its location. But that variability is limited. Doing weight and balance calculations ensures your CG is "within the envelope."

You may be intimidated by all these calculations. Don't be. Like many things in flight training, understanding the background of weight and balance will help you be a better pilot, but you won't use that background and your calculation skills every single time you fly. This is because you will probably fly the same aircraft with consistent loading once you have your PPL and therefore won't have to calculate weight and balance for every flight. You'll come to know the common loads you carry and know that they are within limits. For example, I know that if it's just my wife and me in the plane I own and we have full tanks of gas, we're just fine. But if I'm adding a big suitcase to the baggage compartment, I'll calculate weight and balance to be sure I'm still within limits. Anytime my passenger is somebody other than my wife, I also do the calculations. More to the point, with electronic flight bags like ForeFlight, you can set up profiles for your specific aircraft and enter weights for different positions in the plane, allowing you to instantly see if you are within the plane's acceptable weight and CG limits. Given how easy it is to do, there's no excuse for not doing it!

Chapter 8
Checklists: Cool Pilots Use Them

On May 31, 2014, a Gulfstream IV business jet taxied to the active runway at what would later become my home airport—Hanscom Field. Onboard were two crew, one attendant, and four passengers. One of the passengers was the co-owner of the *Philadelphia Inquirer* newspaper.

At 9:39 pm local time, the airplane rolled onto Runway 11 after being cleared for takeoff. The pilots pushed the throttles forward. The plane gained speed. At about 3,000 feet down the 7,000-foot runway and at about 126 knots, the copilot said, "Rotate," indicating to the pilot it was time to pull back on the control column and guide the aircraft into the air. But the plane didn't get airborne, even as it continued to accelerate. With just 1,000 feet of runway left, the plane was still on the ground and doing 162 knots (186 mph). The pilots were unable to stop the Gulfstream. It exited the runway into a field beyond the runway runover area before crashing into a line of trees and a gully. The plane caught fire. Nobody survived.

The crash was a perplexing one. Airplanes with a lot of airspeed and a little back pressure on the yoke usually *want* to fly.

The wing can't help but do it … the laws of physics work and are consistent. What happened?

Before answering that question, let me tell the story of another crash. In 1935, Boeing was competing for a contract to deliver a long-range bomber to the Army. Their entry into the competition was what they called the Model 299. During a demonstration flight late in the selection process, the bomber prototype rolled down the runway, became airborne, but then stalled at about 200 feet. Way too low to recover, the plane crashed and burned. Two of the five people onboard died. Fortunately for Boeing, the tragedy didn't prevent them from eventually being awarded a contract for the airplane. It became the workhorse of the U.S. bomber fleet and was known as the B-17 Flying Fortress, the aircraft my grandfather flew in during World War II.

I've told these two crash stories because they were caused by the *exact* same thing, demonstrating that it's very hard to keep history from repeating itself in aviation. In both cases, control locks (which keep the flight controls from banging around in the wind when the plane is parked) were left in place when the pilots initiated their takeoffs. While the mechanisms for locking the controls were very different in each instance, both their purpose and the effect of trying to take off with them engaged were identical. The only difference is that the B-17 was able to get airborne, the Gulfstream wasn't.

The B-17 crash woke up a growing aviation community to the risks related to flying more complex aircraft without some reference to standard procedures. The checklist was born.

The unfortunate crash of the B-17 prototype perhaps can be forgiven. The pilots, after all, didn't have a checklist. But the crash of the Gulfstream IV is a different case. They did have a checklist. And I'm certain something like "ensure flight controls are free and correct" is listed on it—and probably more than once. In fact, the checklist for the aircraft I flew at my flight school required a flight control check twice before taking the runway: once before engine start and then again right before takeoff.

While writing this book, Dale "Snort" Snodgrass, the man with more flight hours in the U.S. Navy's F-14 Tomcat than any other human and a renowned airshow pilot, was killed in the crash of an aircraft he owned when it nosed up during takeoff and stalled before cratering into the ground in a classic stall-spin scenario. The cause? You have probably guessed: failure to remove a gust lock before takeoff. It was noted by those who knew him that he was a details guy and was methodical in his preflight procedures. But there's evidence to suggest on the day of the crash he was in a hurry. Not taking the few seconds to rotate the stick in his aircraft in a circle to ensure that the flight controls were free and correct cost him his life.

I sound critical here. And to some degree I am. But as they say, "Pride goeth before the fall." The sad truth is that even when the checklist is in my hand and I'm working through it, I've found it very easy to miss a step. I sometimes fly an Archer III, an aircraft that doesn't have a key to select magnetos and engage the starter but overhead switches instead. I missed the checklist item that instructed me to turn on the left magneto before initiating an engine start; I cranked the starter with both magnetos—which supply energy to the spark plugs—off. Even if you're very early in your flight training journey, you probably know this won't work.

So, while checklists are critical to use, they aren't foolproof, because you aren't foolproof.

Walk-arounds and flows

As you get more comfortable flying, there will be an inevitable desire to simply walk through the items on the checklist without *really* looking at them. This has much to do with the ritual of doing something the same way every time. In fact, it's a bit ironic to consider that repeating the checklist over and over can make it more likely that you won't want to use it at some point. For example, the preflight of an airplane is often called a "walk-around." It's not surprising why: you literally start at one place outside the aircraft and work your way around it until you are back where you started, checking items along the way. But after many, many run-throughs of the items on the checklist, the visual cues of the walk-around itself quickly take over and remind you of what to check. When I step off the wing onto the tarmac when exiting a Piper Cherokee on the right side of the plane (where the only door on the airplane is located), the first thing I'll notice as I look at the wing is the right flap. And, not surprisingly, the first checklist item after checking the general condition of the wing is "Flap . . . Check." After scrutinizing the flap's condition and its linkages, I continue my walk along the trailing edge of the wing, coming next to the right aileron. I ensure it moves freely and its control rod and hinges look normal. Then to the wingtip. I continue this walk to the front of the right wing, the nose of the airplane, the left wing, and the empennage before ending up behind the right wing, right where I started.[30]

---

[30] The empennage refers to the back part of an aircraft where the tail and horizontal stabilizer are. The word is derived from the French *empenner* which means "to feather an arrow." So, while the word may not be familiar to those outside of aviation, it describes perfectly what it refers to. The horizontal and

This stroll around the outside of the plane mimics the order of items on the checklist—or, more accurately, the checklist was written to match a logical flow around the plane.

Knowing a flow, even without reference in the moment to a checklist, is a beautiful thing. But it should never be done without a cross-reference back to the checklist at some point. I'm to the point now where, after doing a complete walk-around without looking at the checklist, I stand in front of the plane right in front of the spinner and go through each item to ensure that I covered everything, starting with that "Flap . . . Check" item.

Flows aren't just good for the preflight, they are exceptionally useful while in flight. You should in fact spend some time learning flows related to the many emergency procedures for your airplane. This is because some emergencies need immediate attention, so you won't always have the luxury of pulling out a collection of checklists, finding the right one, and then methodically stepping through it.

There were two things I found very helpful when trying to get emergency checklists committed to memory as part of a flow. First, I found a schematic of my airplane's controls and instrument panel on the internet and printed it out. I then drew numbered arrows from one control and/or instrument to the next to show their relative order of steps to undertake in a critical situation. For example, the "Engine Fire in Flight" checklist has five things to do immediately:

1. Fuel Selector . . . OFF
2. Throttle . . . CLOSED

---

vertical stabilizers serve the same function as the feathers on an arrow: to keep the aircraft pointed in the direction of travel in a controlled manner.

3. Mixture . . . CUTOFF
4. Fuel Pump . . . OFF
5. Heater and Defroster . . . OFF

On my schematic of the cockpit, I have arrows that run from the lower left of the instrument panel (where the fuel selector valve is located) to the throttle in the center console, to the mixture (also in the center console), to the fuel pump switch, to the heater and defroster levers. I then committed to memory the last two steps after these five: 1) initiate an emergency descent with speeds around 120 knots to get down quickly and (hopefully) put out the fire, and 2) prepare for an off-field landing. That diagram of my cockpit with arrows overlaying it was a visual that was easy to recall in my mind's eye, helping me to know where to move my hands as I executed the checklist steps without even looking at it. Again, though, time permitting you should always backup your flow by reference to the checklist after the fact.

The second thing I did when it came to emergency procedure flows was spend some time not just memorizing the steps but thinking through *why* the steps are necessary. In the case of an engine fire, it doesn't take a firefighter to understand that your priority should be to starve the fire of its fuel source. The best way to do this in most aircraft is to cut off the fuel flow from the tanks to the engine. Additionally, closing the throttle and cutting off the mixture ensures that any fuel that has already made it past the shutoff valve can't be mixed with air in the engine, eliminating another potential fire source. Turning the fuel pump off keeps fuel from being delivered to the engine. Lastly, turning off the heater and defroster ensures that any smoke from the engine compartment doesn't enter the cockpit.

The Invisible Gorilla

Even with checklists, flows, and mnemonics though, it is unbelievably easy to forget items on a checklist or forget the checklist altogether. In thinking through why this might be, I'm reminded of a famous experiment carried out at Harvard. In it, a group of subjects was shown a video of a ball being passed between six people arranged in a circle. Three of the ball-passers wore white shirts and three wore black shirts. Subjects watching the video were asked to count the number of passes between those in the video wearing white shirts. Easy enough, right? Well … yes, it is. But here's the kicker: during the ball-passing, somebody wearing a gorilla suit walked into the middle of the circle, faced the camera, pounded their chest, and walked off the screen. When the ball-passing came to an end, those watching the video were asked if they saw the gorilla. Half of those who participated did *not* see the gorilla—which has since become known as the "invisible gorilla."

How in the name of all that is holy is that possible? It's truly perplexing. Search for the video online and you can find it. You'll have no problem counting the passes *and* seeing the gorilla but, of course, you know about the gorilla ahead of time. The phenomenon exhibited by those who watched the video and didn't see the gorilla while counting the passes is called "inattentional blindness." In a paper written in *Psychological Science*, in 2014, the three authors (Drew, Vo, and Wolfe) define inattentional blindness as missing salient events in our world if we are engaged in a different task.

Flying is an activity where we are *always* involved in a "different task." By this I mean you are almost always managing multiple workstreams. And when you are doing that, it can be easy to forget to refer to a checklist, employ a mnemonic you have learned, or check off a checklist item. I know … it's happened to me. While building hours as a solo student pilot, I was approaching Hanscom for landing and had been given instructions telling me to enter a left downwind for Runway 29, the approach I was most familiar with as this was the runway used probably 80 percent of the time given prevailing winds. As usual, the airspace was very crowded with aircraft like mine—small, slow piston-driven planes and faster, heavier business jets. I set up for a 45° entry to the downwind leg of the pattern as I was trained to do.

Just as I was about to come abreast of the numbers on the far end of the runway (i.e., the side opposite the end where I was going to set up for a final approach), the tower decided to change the landing runway to Runway 11. Given the complicated choreography required to manipulate a crowded traffic pattern into a new flow for a new runway, the tower asked me if I could do a short approach so that I could get on the ground quickly and get out of their way (they didn't say this last part but in listening to the radio traffic it was clear that this was what they had in mind). I agreed, having done several short approaches during my training. For the uninitiated, a short approach is one where you have an unusually short final because you are turning base when you are essentially abeam the end of the runway. I immediately made a left turn and aimed just to the left of the threshold, putting me on a 90° intercept angle for the runway centerline. I had little time to get ready for landing. I put in some flaps to slow down and began a turn to a final that would have me rolling out of my turn on the centerline just as I was crossing the threshold of the runway. I got full flaps in and then greased the landing (as far as any of you know). I got off the runway as soon as possible, knowing that there was a Gulfstream conga line behind me.

When I exited the active runway, I pulled out my checklist and proceeded to step through the after-landing items. When I got to the fifth item—Fuel Pump . . . OFF—I realized it was *already* off. And that's because I had never turned it on during my setup for landing. It should have been turned on for redundancy. The airplane had an engine-driven fuel pump so turning on the electric pump during landing was a way to ensure that, if the engine-driven pump failed close to the ground, I wouldn't find my engine quitting from fuel starvation.

While all turned out okay, in my own mental debrief of how I forgot to turn on the fuel pump, it became immediately clear how I had screwed things up.

Like many pilots, I use the BC-GUMPS mnemonic to set up for a landing. While there are mild variations on what each of the letters stand for, here's how I define them:

- B—Boost Pump; otherwise known as the fuel pump, on.
- C—Carburetor Heat; configured on or off depending on the aircraft.
- G—Gas; ensure you have selected a gas tank with gas.
- U—Undercarriage; is the landing gear down?
- M—Mixture; it should be full rich in most cases to ensure max power if you need it on a go-around.
- P—Propeller; this one is meant for those flying aircraft with constant speed propellers. Since I fly airplanes with a fixed pitch propeller, I instead let the "P" stand for "power"—I like to ensure the engine RPM is where I want it to be.

- S—Safety and Switches; are your seatbelts fastened? Landing light on?

I would usually conduct my BC-GUMPS checklist when I was midfield on my downwind. But the sudden change of runway and the sense that the controller was busy and really needed me on the ground to help sequence in faster aircraft made me behave like somebody counting the number of passes between the white shirts in the invisible gorilla experiment. I got so focused on one thing—completing a short approach to comply with the controller's instructions—that I forgot my BC-GUMPS checklist even though there was ample time. It was my invisible gorilla in that moment. There was another factor that contributed to my inattentiveness. While, as I mentioned, I had completed several "short" approaches during my training, I hadn't done them while also changing from one runway to another. The views were different, pattern abnormal, and there was a LOT of chatter on the tower frequency. All conspired to distract me from my normal flow.

Don't be a deviant

The thing that's tricky about checklists is that not only can they be easy to miss when you are distracted, but they can also be easy to miss when you aren't distracted at all. The crash of the Gulfstream at Hanscom I discussed earlier was a "normal" takeoff by all accounts. But by normal, I don't mean that everything was done by the book. Rather, I mean that the crew did what they had apparently done repeatedly in the past in a rote way. In the NTSB investigation into the accident, it was discovered using data from a flight recorder that the pilots hadn't performed complete flight control checks on 98% of their previous 175 takeoffs. They weren't distracted at all; they had just made as standard practice an improper behavior that hadn't bitten them before.

There's a name for this. It's called the "normalization of deviance." The term was coined by Diane Vaughan, a sociologist at Columbia University. Her term and the definition of it—the process by which deviance from correct or proper behavior becomes normalized in the corporate culture—became well known because of her book, *The Challenger Launch Decision*. In that book, Vaughan noted that engineers at NASA allowed pre-established safety rules to be violated when they found that violating them had no adverse consequences. With schedule and budget pressures at play, it became easy to implicitly and explicitly rewrite those rules of safety to continue launching shuttles with a known design flaw.

In the case of the Gulfstream crew at Hanscom, this is exactly what happened. The crew had gotten into a habit of not checking their controls before beginning their takeoff. From their perspective, it probably seemed unnecessary. After all, in every case before that deadly day at Hanscom, the takeoff proceeded without event. In fact, it's entirely possible that at some point they simply forgot that it ever was a requirement to check the flight controls . . . although I must admit that this seems incredibly hard to believe—especially since checking flight controls takes virtually no effort at all, so why not do it?. The other peculiar factor in this accident is one that is important for those of you who will be flying without a copilot—in other words, all of you. You'll be required to fly solo before you get your certificate and will probably fly solo even after your checkride. The crew in the Hanscom crash flew together regularly. This seems to have led to a familiarity with each other that quite literally became deadly. I suspect that if the pilot in command when the crash happened had had a different copilot on each of the 175 takeoffs the data recorder captured leading up to the accident, most of them would have asked about checking the flight controls per the checklist. But since the pilots had normalized deviance together, neither one called out the other. It was as if the pilot in command was in the cockpit without a copilot—just as you will almost always be. What happened to this crew is relevant for you as a single pilot: you don't have somebody to challenge *you*.

So how do you ensure that you adhere to checklists and don't slip into your own deviant normalizations? The best answer I've found is to—without fail—read your checklist items out loud and tell yourself you've completed them out loud as well … even if you are alone. There is some sort of self-accountability this drives. In effect, you need to, for a moment, have multiple personalities. I call my copilot personality Checklist Charlie. When I'm going through my checklists, I ask my alter ego if he has confirmed or completed the items I'm reading. I like to imagine my checklist buddy has a British accent like Lord Haldane but more youthful and without the patina of arrogance I imagine he might have had in his voice. Checklist Charlie's accent is laced with cheer and enthusiasm.

Me: "Why my good fellow, Charlie Checklist, did you confirm that carb heat is off?"

Checklist Charlie: "Well I most certainly did Patrick."

Okay, I don't use those words *exactly*. But I do read the checklist items and respond to them with my own voice. If I'm with passengers, I warn them that I'll be doing a lot of talking to myself. This helps put them at ease (I hope) and gives them a clue as to when the cockpit should be "sterile"—i.e., free of distractions like unnecessary conversation. I tell them that if I'm doing a lot of talking, they shouldn't be.

The callouts to myself are accompanied by hand and eye movements that check the status of something or execute an action. This statement may seem like a "no duh" one: of course you have to move your eyes and hands to, for example, turn on your master switch. But it turns out this connection of verbal verification to motor skills is something that those who have studied checklists say is very important. In a paper by researchers Asaf Degani and Earl Wiener called "Cockpit Checklists: Concepts, Design, and Use," the authors note the following:

> *Besides visual verification of the check item, motor movements such as touching controls and displays ("muscle memory" as some define it), are also an effective enhancement for the verification process. The use of the hand to guide the eye while conducting the check can substantially aid the checklist procedure by combining the mental sequencing process with motor movements.*

I agree. In fact, I like to think that the combination of my own voice, the movement of my hands to execute the checklist item, and the movement of my eyes from the checklist to where my hands are going constitute a sort of three-channel system that is constantly checking itself. As you become familiar with your checklists, you'll feel this connection between the three legs of your checklist execution system take hold. In fact, it will become awkward if any one of the three things don't match for some reason. You'll notice this if you fly one aircraft a lot and then transition into a new one where the location of, for example, the avionics power switch is not where you expected. You'll read the item that tells you to turn it on, move your hand in the direction you expect the switch to be, and then look to find it's not there. This is your "does not compute" moment and should force you to pause and consider what went wrong.

Make the checklist work for *you*

Another thing I found helpful was to create my own checklist. To be clear, my checklist had all the same steps in the same order as the checklist the school provided to me and as the POH recommended. Why, then, create a new one? There are three reasons. First, there was formatting on the provided checklist that was meant to emphasize certain important steps but, for whatever reason, made it *more* likely I'd miss the step. It's almost as if the most prevalent formatting locked my brain into looking for it and it alone at the exclusion of formatting meant to emphasize a point; the highlighted items, in some cases, became invisible to me. After several times missing such items, I decided to format it the way I wanted. Not only was the new checklist clearer for me to read, the simple act of changing it to my preferences put a bookmark in my brain related to the step; I was more likely to remember that I had changed a given step in some way and therefore less likely to miss the step.

The second reason to modify a checklist is to make some steps more discreet. By this I mean that some checklists include two steps combined into one. On my school's checklist for the Piper Warrior III, the "Engine Runup" section includes this:

LOAD METER, ALL LIGHTS, FUEL PUMP, & PITOT ON .... CHECK THEN OFF.

This is a check to ensure that the alternator's output can match the demand of heavy electrical load before takeoff. I know it seems silly to many of you, but I often would read the first part telling me to turn on all the electrical systems and then forget the second part telling me to turn off what I had turned on. This was solved by creating two distinct steps: one to turn on the equipment and one to turn it off.

Finally, often checklists would be better if things were added to them. Some items you may add are directly related to the safe operation of the aircraft. But additions can also be much more pedestrian. I fly with a ForeFlight Sentry ADS-B device which, when connected to an iPad, gives me real-time data related to my position and other air traffic (among other things).[31] The Sentry is about the size of a deck of cards and is usually mounted to a window on the airplane with a suction cup. I have had a terrible habit of forgetting to retrieve my Sentry at the end of a flight, only to return home and wonder where the expensive device was. I've done it twice! To the credit of the East Coast Aero Club pilot community, when I called the club about a week later and asked them to check if the Sentry was where I had left it, it was. In any case, I added a "Retrieve Sentry" item to the end of my checklist.

---

[31] We will discuss ADS-B in more detail later. But the acronym stands for "Automatic Dependent Surveillance—Broadcast." To simplify greatly, ADS-B is a system that allows aircraft to transmit their position and other critical information so that both ATC and other aircraft can have a clear picture of the activity in an airspace.

Chapter 9
Can We *Go* Somewhere?

As the winter turned into spring in the Boston area, my flying increased in frequency. I tried to get two flights a week in to help what I'd learned "stick" from one lesson to the next. That said, frustrations abounded. I've since learned that the frustrations I had are common among those who embark on the private pilot journey.

Scheduling and weather

The first frustration was getting flights on the calendar. My school had a great online scheduling system that allowed me to reserve an airplane and my instructor up to 30 days in advance. The problem was that to get a slot you had to plan to access the system in the same way you might need to plan to buy tickets to Taylor Swift concert on the day tickets went on sale. I'd log in every few days to schedule out the full 30 days before all my classmates at the school gobbled up any availability. Finding a slot where both my instructor and an airplane were available was a challenge, and that doesn't even consider the need to deconflict with my work schedule and family obligations. It can become maddening to find a time and day when all the scheduling stars would align.

After finding a way to get training times on the calendar though, there was still something else to worry about. And as strange as it seems, this other "something" never crossed my mind before I began training: the stupid, unpredictable, frustrating, cursed, evil, unfriendly, uncooperative, devil-sent weather. In more than half the cases, all my Rubik's Cube-like schedule planning was for naught. Because, in the end, the weather got in the way.

I'll spend some time later talking more about aviation weather but suffice it to say that unless you live in the deserts of the southwest United States or parts of the Australian outback, this will also be your regular nemesis. In §91.155 of the Code of Federal Regulations (which you find in your *FAR/AIM* book), paragraph (c) notes that "no person may operate an aircraft beneath the ceiling under VFR within the lateral boundaries of controlled airspace designated to the surface for an airport when the ceiling is less than 1,000 feet." The regulations also say that visibility in most cases should be no less than three statute miles. But like a lot of FAA rules, these are absolute minimums that can entice you into believing that because you're legal, you'd be smart to fly. You wouldn't be.

One thousand feet of vertical space to work in is not very much. It puts you at risk of all sorts of bad outcomes like running into an obstacle or flying into clouds when the ceiling creeps lower, risking spatial disorientation and loss of control. It also doesn't leave you a lot of altitude to buy yourself time and options if you have an engine issue. The other problem you have is that most traffic patterns for light aircraft are at 1,000 feet above ground level (AGL).[32] But those same VFR flight rules that allow you to fly if the ceiling is at 1,000 feet also say that you must stay 500 feet below clouds in many airspaces. So that means that if you take off from your controlled airfield with 1,000-foot ceilings, you'd have to fly at 500 feet. That's way too low to be safe in or out of the traffic pattern. I didn't like to fly away from the airport unless ceilings were forecast to be at least 4,000 feet. I was okay with flying in the pattern with a ceiling lower than that, knowing I'd always have the airport in view, say 1,700 AGL. But even then, I'd be very careful to monitor the conditions. With any signs of them worsening, I'd get my butt on the ground.

Even with no ceiling, your flying can become dangerous if visibilities are low. Like the 1,000-foot ceiling rule, the three miles visibility one seems ridiculously unrestrictive to me. I experienced firsthand how tough visibility below about six miles can be to fly in. I was flying one humid summer day at Hanscom. I was just doing some pattern work and visibility was forecast to be over six miles. However, smoke from western wildfires had been caught in the jet stream and the view out the window deteriorated quickly. Eventually, spotting other traffic in the pattern and locating the runway if I was sent on an extended downwind, for example, became difficult at best.

---

[32] AGL is literally how high you are above the ground. It's (almost always) different than MSL (mean sea level). MSL is elevation as read on a correctly configured altimeter (i.e., one corrected for local atmospheric pressure). MSL altitude is usually *not* telling you how high you are above the ground. But it does help ensure that all aircraft are using the same altitude reference so that if two aircraft near each other are flying at 3,000 feet MSL, they will be at the same height above the ground, even if that height is actually just 2,500 feet. I'll discuss the altimeter and how it works later in the book.

The FAA defines visibility as the distance at which you can "see and identify prominent unlighted objects by day and prominent lighted object by night." Many who haven't experienced this limited visibility probably can't appreciate just how scary it can be. I think that our terrestrially bound brains think of three miles as a long way. If you put me on perfectly flat ground and pointed at a house on the horizon and said, "That's three miles away," I'd think that visibility was pretty good. But once you get into three dimensions and start moving fast, three miles doesn't seem far at all.

To understand this better, I pulled up a satellite image of Hanscom field and plotted the place where I typically make my turn to base for Runway 29 assuming I was flying a standard pattern. Adjusting for my altitude since the measurement I referenced in the last sentence is a distance over the ground, I determined that the end of the runway was about 1.4 miles from my location in the air. I also calculated that the far end of the runway was about 2.42 nautical miles from that base turn location. You may think, "Well, if you have three miles of visibility, you should be able to see about 0.6 miles beyond the far end of the runway." But note that the visibility is not discreet in nature, meaning that it's not like it's completely clear all the way to three miles *and then* it gets murky. No, it's a continuum. Everything, no matter how far away, is harder to see than if it was "clear and a million." This is very important to keep in mind. I won't fly away from the pattern if I can't be reasonably sure that visibility will be at least six miles and preferably ten.

And, of course, there's the wind. High and/or gusty winds, especially if not aligned well with runways you might be landing on, can lead to frequent cancellations in many parts of the world. One thing I learned during my training though—and it's been confirmed for me since getting my certificate—is that wind is something that many pilots can be *too* conservative about. If you wait for a dead-calm day, you may never fly. Make it a point to fly in (appropriately) strong winds when you are with an instructor. This will help you gain proficiency, something sitting on the ground every time a breeze blows won't do.

Boredom

The other problem I ran into as I entered the middle part of my training was something many would describe as boredom, but I think is better described as training fatigue. Many instructors, especially at a busy school, will focus on getting through the training, which means a focus on the *ACS* and all its requirements. Jeff and I would take off from Hanscom and fly out to the practice area to repeatedly go through stalls, steep turns, turns around a point, and all the other things required in the *ACS*. After a while, I felt like a young military man who *only* trained and never fought. I was losing the connection between what I was learning and all the wonderful aspects of being a private pilot I'd enjoy once I had my PPL. While I never voiced my fatigue with Jeff, I wished I had. I don't think it would have been inappropriate to say, "Jeff, today can we just fly somewhere and get lunch?"

In retrospect, I'm sure he would have *loved* to hear this from me. When I was wrapping up my training with my new instructor, I mentioned to him the frustration I had earlier in my student pilot days and my idea of just flying somewhere to eat. His eyes lit up as if to say, "I wish you'd asked me to do that!"

And then, it hit me.

Of course, I thought.

To the extent that a student has fatigue during their training, an instructor has it in spades. After all, I only went up, on average, less than two hours a week when all was said and done. But an instructor could fly eight or more hours a day. No matter how much you love to teach and fly, doing the same thing repeatedly gets old. Who wouldn't want a break from that monotony?

It's common for students to feel the way I did. You likely will too. When you do, don't be afraid to ask to do something that would give you a hint of what you could do once you have your PPL. Fly to get lunch or breakfast. Fly to a landmark you'd like to see from the air. Fly to a different airport and attempt a landing and then get out and stretch your legs. Do anything that brings joy back to aviation and reminds you of why you started on the journey in the first place.

Measuring yourself

Finally, there is the issue of measuring yourself against others. This one *really* surprised me; like the vagaries of the weather, it was a totally unexpected frustration when I began training, that, in retrospect, should have been as obvious as the sun in the sky on a (seemingly rare) sunny day in Boston.

In virtually all training and learning environments we experience in other dimensions of our lives, we typically are part of a group and therefore have a sense of our performance relative to others in that group. For example, throughout all levels of formal education—primary school, high school, and university—we are engaged in learning directly with other students, allowing us to get a sense of each other's relative strengths and weaknesses. Additionally, because of things like class rankings, grades, standardized tests, and other measures, you can have some idea of whether you are on one end of the bell curve or the other. This is true in many other settings as well. If you've ever played organized sports, you know how your skills stack up against the skills of others—performance on the field or court makes this hard to hide. In your job, performance evaluations and promotions usually make clear how good you are relative to those you work with. If you run your own business, revenue and profit relative to the competition provide a concrete measure of (at least one kind of) success.

But learning to fly, at least at a Part 61 flight school, is not like this at all.[33] Sure, there were plenty of student pilots in a sort of loosely organized learning community at East Coast Aero Club. But none of us knew how others were doing in their training. There were no dogfights allowing us to count our "kills" or forced-curve performance rankings of the kind Maverick and Goose had in *Top Gun.* Each student was on a sort of learning island with their instructor.

That's not to say that my instructors didn't give me feedback. They did. But most of it was tactical in nature, related to something happening in the moment.

You should turn the yoke to the right while taxing given where the wind is coming from.

---

[33] "Part 61" refers to the same federal regulations I discussed earlier. In that code, the rules are laid out for flight schools that allow a more flexible training environment. Most recreational pilots will take this path. It's a better one for part-time students who can't be tied to a "set in stone" schedule. The school I used for my training was a Part 61 school. This contrasts with "Part 141" schools which have more structure, are more oriented to full-time students, and—if they've been approved for it—allow students to get their certificates in fewer hours of flying than Part 61 schools. For these reasons, Part 141 schools are typically appealing to those who want to get training done and move on to a flying career with the airlines or other commercial carriers.

Don't take your hand off the throttle during takeoff.

Keep your toes off the brakes unless you need to brake.

I'm ashamed to say it, but I needed more. I wanted to know that among the one hundred or so students who started around the same time I did and flew about as often as I did, that I was in the top 10% of student pilots. (I know what you're thinking … I could have been in the bottom 10%. But absent any objective measure, and as long as I'm the one doing the writing, I choose to be more "optimistic" in my imaginary ranking.)

Undoubtedly this desire to know my place among many has an ego aspect to it. But there were three other reasons I craved to know where I ranked in the larger student pilot population. First, it would have helped give me comfort that I wasn't the only one struggling in some areas, a valuable bit of knowledge when in the training doldrums. Misery does indeed love company. Secondly, to the extent I *was* "ahead of the class" in some areas, it would have been helpful to know more precisely where I lagged. This would have allowed me to focus my training more efficiently. Of course, a CFI should have a sense of this and direct you accordingly. But that leads to my last point: instructors are human and have their biases. I certainly think it's possible that they could give high marks to a student who's marginal but whom they personally like while giving less favorable feedback to somebody they dislike but who is the world's next Chuck Yeager. Knowing where I ranked in a group would have given me a more objective assessment of my skills and progress.

All that said, the fact of the matter is that a ranking among a cadre of other student pilots is not in the cards for most of us. In that case, is there anything you can do? I think so. I would suggest speaking openly with your instructor at the beginning of your relationship to ask them how they will signal to you your progress against the standards. I believe many instructors think they do this, but in my conversations with others who have gone through pilot training at a Part 61 school, I don't think they really do. My suspicion is that because most instructors do a lot of instructing, they can fall victim to the bias of believing that what they see as progress, success, or failure is obvious to the student. Often though, that's not true. To guard against this, be explicit in your requests for feedback. Don't settle for general comments. In a post-lesson debrief, don't let your CFI get away with saying something like, "You were too fast on final." That's an interesting piece of information, but not nearly as valuable as if they said, "You were late in reducing power on the downwind and also late deploying your first notch of flaps, so by the time you made your turn to final, you had too much energy."

Also, after each lesson, make notes about what you did and how well you think you executed whatever you did, recording a combination of your instructor's thoughts and your own. It might look something like this (continuing our example from above):

Date	Maneuvers	CFI Thoughts	My thoughts	Conditions
3/5	Pattern work, Touch-and-Gos	RPM not right in downwind, too fast as we approached base turn, flaps in late, 10 kts too fast on short final led to a lot of float, touched down well past 1000 ft. markers	Distracted by a call from the tower and was looking for traffic. Didn't manage power and was well beyond the numbers when I put in flaps – put them in right before base but didn't reduce power first, too much energy as I crossed runway threshold	Visibility 10 SM+, winds 260 @ 8-10 kts pattern crowded, Runway 29 in use

It's important to capture both your own thoughts and those of your CFI because, as the example shows, there are things you may have been aware of during a given maneuver that your instructor wasn't. And those insights can be crucial to learning. In the example, the instructor knows I was late in taking out power on the downwind but didn't know in the moment what I recorded in the "My Thoughts" section: that I was distracted by a call from the tower asking me if I had landing traffic in sight. This piece of information may be an indication that you haven't yet mastered the multitasking that flying requires and prompt a discussion with your instructor about how best to remedy that. For example, you and your instructor may agree that he or she will help with the radios until you have the cadence of the pattern nailed along with your landings. As you get more practice, you'll record further entries in your flight diary that will help you see if you've made progress against this "better multitasking" goal, ultimately helping you to know that you are meeting the standards laid out in the *ACS*.

To the last point about the standards, I want to be clear that while I missed the ability to compare myself to others and think there would have been utility in being able to make that comparison, in the end meeting the requirements laid out in the *Airman Certification Standards* is what is necessary. So, find your own way to keep score and hold yourself accountable by reviewing the standards frequently and noting how you are performing against them.

A quick note on scorekeeping

At the risk of contradicting myself, I'd caution you not to get too fixated on one insidious form of scorekeeping: counting hours to major milestones like solo flights and passing checkrides. I've seen numerous questions posted online in various student pilot groups where the writer is asking how long it took others to get to their first solo flight or a successful checkride. Answers to this question are, for the most part, meaningless.

I soloed in about 19 hours. If I believe some of the noise on the internet, that number makes me an abject failure. But that number (for me or for you) could have been higher or lower depending on many factors. For example, the frequency with which you're able to fly is critically important. Cognitive and muscle memory atrophy the longer you go between flights. If you fly infrequently, each hop into the cockpit will require refreshing yourself on things you learned on your previous flight. This takes more hours. The weather can play a role too. You may fly with an instructor in winds that they wouldn't allow you to fly in by yourself. You may be ready to solo, but the weather isn't permitting it. Meanwhile, your total hours keep clicking upward. You may just have a more conservative instructor than other students. And, yes, it also may be true that it's just taking you longer to "get it." All of this is normal.

So, don't get into the comparison game *too* much. That said, if you're 40 hours in, for example, and haven't soloed, ask your CFI to speak frankly with you about what is holding you back. Give them permission to be brutally honest. If their answer doesn't satisfy you, then seek out another opinion. There's no law against flying with two or more CFIs. In fact, I found it helpful to do a couple of flights with somebody other than my primary CFI. This helped me see other learning techniques, some of which resonated with me better than those of my "main" CFI. And it gave me a chance to hear somebody else's opinion about how I was progressing.

Teaching is hard

While we're on the subject of CFIs, it's important to remember that, like most of the general population, they aren't necessarily gifted teachers. This isn't for lack of trying by the FAA or by them. The process to become a CFI includes learning about ... well, learning. The *Aviation Instructor's Handbook* has more than 200 pages of information to help CFIs become good at teaching. But in the grand scheme of things, it's doubtful to me that somebody who doesn't have a knack or the patience for teaching is going to be turned into a master teacher through the FAA training program.

How do I know this? Teaching well is hard. And you don't have to be an expert on the topic to know this to be true. Think of every teacher you've ever had in a formal education setting. How many of them were world-class at what they did? I'm guessing few. And yet these men and women (your high school and university teachers, for example) had years of training to do what they do. Most of them—at least in the United States—probably had master's degrees. The number of hours they spent studying both learning and teaching is well beyond that of your typical CFI. No, studying how to teach doesn't mean you'll be a great teacher, whether you're teaching sixth grade math or the basics of a traffic pattern.

However, there is one thing most flight instructors have going for them that I don't think my instructors in other settings necessarily did. And it's something that's beneficial to you. They had a burning passion for the thing they are teaching. During my training, I've flown with six CFIs. To a person, each of them *loved* aviation. While passion alone won't guarantee a great learning experience, the enthusiasm that comes with it certainly makes sitting to the left of what might be a middle-of-the-pack teacher much more fun and interesting.

But don't rely on enthusiasm alone. While I didn't do it—and think I got lucky with good instructors—I wouldn't be afraid to "shop around" when you begin your training. In a *Flying* magazine article from June of 2021, Julie Boatman echoes this sentiment, and more. She writes that there are five things you should do to find a great instructor:

1. Interview several instructors before you pick one.
2. Take note of an instructor's use of learning tools, such as a syllabus and good (pre- and post-flight) briefings.
3. Work with somebody who can train on your schedule.
4. Hire instructors who value their time (i.e., they charge for it) as it's likely they will value yours.
5. Look beyond somebody who may "fit" simply because of demographic or other reasons (i.e., don't look for familiarity, look for competence and efficacy).

Number one may be hard to do. If you use a busy Part 61 flight school as I did, the pace of training and the variety of instructors to choose from can make asking for multiple meetings tough. While some schools may let you meet with individual CFIs, keep in mind that if they are sitting with you answering questions, they aren't earning money with their current slate of students. To remove any awkwardness around this, offer to pay for the CFI's time as if you were a student (some schools may require this without you offering). Don't be annoyed by this. CFIs sell their time, and a half-hour conversation won't cost that much anyway—particularly considering what you are preparing to spend to get your certificate.

The other piece of advice from Ms. Boatman's article that resonated with me was probing your CFI about their use of learning tools. At a busy school, there can be an assembly line feel to things. You arrive for your two-hour block and might start a little late as your CFI tries to squeeze in time to grab a bite to eat or hit the restroom between lessons. They then may want to spend a few minutes talking with you about the lesson for the day. Next, you need about 15 to 20 minutes to properly preflight your airplane. You may have to wait for fuel or get to the pumps yourself. Depending on how busy your home airport is, you may then spend another 10 to 15 minutes taxiing and waiting for takeoff clearance. By the time you get in the air, you could already have burned 45 minutes of your time. If you then spend an hour in the air, by the time you land, taxi back to the ramp, and tie down the airplane, you've got barely enough time to do a debrief with your instructor.

Not surprisingly, this frantic pace *will* get in the way of the use of proper learning tools. First, it can be hard for a CFI in this environment to keep track of each student's progress, measuring it against an established syllabus. This is why I recommend keeping your own training diary. Second, and this is related to the first challenge, the schedule crunch stacked lessons introduce makes it harder to find time to do a proper debrief after your flight lesson. This is an especially common thing because I think many instructors are biased in the direction of using your training time for actual flying (vs. ground instruction). This is admirable—after all, you're there to fly!—but a proper debrief is often as important (or more important) than stick-and-rudder time.

The remedy to ensure the proper use of training tools is, as with most things, communication at both the beginning and end of a lesson. Here's a short checklist for you to use to ensure you keep that communication happening from lesson to lesson:

When you arrive for a lesson

1. Remind the instructor of what you did during your previous lesson, referencing your flight diary.
2. Ask them what the plan is for the current lesson.
3. To the extent that the current lesson involves maneuvers you haven't done before, ask them to review the proper setup and execution of the maneuver and the standards that must be met to consider the maneuver successful. Don't wait until you're in the air to get an understanding of the setup for a maneuver and execution of it.

When the lesson is done

1. Ask for *specific* feedback regarding how well you executed on the goals for the day.
2. Document what you learn during this debrief in your diary.
3. Ask them for homework related both to the lesson just completed and what they anticipate you will do during your next lesson. Examples of homework might include reviewing the steps for practicing power-on and power-off stalls, using a flight simulator to practice instrument scans, or "chair flying"—the practice of visualizing maneuvers and moving your hands in unison with checklist callouts

to help you memorize activities necessary during critical phases of flight. (When you chair fly, be sure to do the movements in real time, as you would when in the air. As they say, practice like you play.)

Doing these six steps religiously will help to get you into a sort of learning flywheel. Many student pilots tend to think that learning to fly is all about the stick-and-rudder part of learning—the actual flying. But your brain is an amazing thing and frequent intellectual engagement outside the cockpit accelerates learning which in turn accelerates proficiency. Listen to aviation podcasts while commuting. Visit the AOPA website (www.aopa.org) and read the amazing content there. Use a flight simulator. These out-of-the-cockpit activities will make the most of your time in the cockpit.

Chapter 10
The Written Test

At some point during the early part of your training to become a pilot, you will need to begin preparing for the written exam the FAA requires you to pass before earning your certificate, the Private Pilot Knowledge Test. For many, this hurdle on your way to your wings is the most daunting. "I'm not a good test taker" is a frequent refrain I've read on student pilot Facebook groups when a member posts disheartening information about either failing the exam or getting a barely passing score. There is no doubt that some of you will be better at internalizing the information necessary to pass the exam than others. And that may seem unfair. After all, it's certainly true that you could be an amazing stick-and-rudder pilot without having the FAA's "book" knowledge crammed into your cranium. But I'm not one who believes that the exam is superfluous. Some level of book knowledge about the many facets of an activity that can kill you or others is a fair requirement, I think. More importantly, I'd posit that if taking a standardized exam—one where you generally know what you are going to be asked ahead of time—is anxiety inducing to such an extent that it doesn't allow you to perform well, then flying may not be for you.

To that last point, while the FAA doesn't state it explicitly, I think the exam isn't just about testing your knowledge; it's also about requiring you to perform under pressure. Because there is *always* some form of pressure when flying. Those pressures range from trying to listen to ATIS on one radio while also keeping an ear out for your call sign on another to a catastrophic engine failure. Handling either of these situations, and the millions of things between the two extremes, in a measured but decisive way is critical if you are to be a safe pilot. Passing the exam is a small, but important, achievement to show you can perform when the clock is on you and the stakes are high.

The latter opinion is probably not too controversial—although I'm sure some will even fight me on the point. But this one might be more controversial: I think the FAA is *too* generous when it comes both to the score you can get and the number of times you may take the exam.

As of early 2023, the exam has 65 questions but only 60 of them are graded; the other five are questions the FAA is testing for inclusion in future exams. You won't know which questions are graded and which are not, but you must get 70% of the 60 graded questions correct to pass. That means you may get 18 questions wrong and still pass. In early 2023 the FAA also reduced the time allotted to take the exam from 150 minutes to 120 minutes. Even this shorter time limit is plenty of time; many I've spoken too took no more than about 90 minutes to complete the test. You may take the test as many times as necessary to pass. In a recent year, nearly 32,000 people took the exam and 90% passed it; the average score was 83%. That 90% figure includes those who took it more than once. Seventy-eight percent of first-time takers passed it. I admit the high scores may be partly due to the motivation of those who embark on earning their PPL. We are generally a high achieving lot so the pool of those who ultimately take the exam is not representative of the general population. To make this point: research scientist Jon Miller of the University of Michigan (and formerly of Michigan State University) concluded that "70% of Americans cannot read and understand the science section of the *New York Times.*" He also noted that only 28% of Americans qualify as "scientifically literate." My suspicion is that those seeking a PPL don't fall into those inauspicious groups.

There are some who will disagree vehemently with me not just regarding what I believe is a low threshold to pass the test, but regarding the utility of the test in the first place. In an article in *Flying* in 2022 titled "Should Knowledge Tests Before the Check Ride Go Away?" writer Michael Wildes expresses his skepticism regarding the use of the exam at all. He speaks for many.

In a major thrust of his argument, he highlights initial checkride pass rates and written test pass rates in 2020 to make this point:

> *The interesting thing to note is the higher pass rate and average score of the knowledge tests [83% pass rate] compared to the initial check ride pass/fail [77% pass rate] statistics. Is it sufficient to continue making the case that knowledge test scores are a good measure of whether or not someone will do well on their checkride?*

He seems to argue the higher pass rate for the written exam shows that it does not adequately prepare pilots for the checkride because the pass rate on the checkride is lower. It's as if he believes that if 83% of written test takers pass the exam then at least 83% should pass the checkride. But this doesn't consider that the checkride pass rate might be *even lower* if the written exam wasn't part of the process.

The other, more salient problem with this argument is that, as far as I know, the FAA doesn't even make the case that the purpose of the written test is to explicitly assist you in passing your checkride. In fact, I'd argue that the two forms of testing are required parts of the process because they take care of two slightly overlapping but mostly distinct requirements.

Yes, it's true that the designated pilot examiner (DPE) who will administer your checkride will review with you areas related to questions you missed on your written exam.[34] But even for an individual who scored the minimum 70% passing grade, the review of those areas during your checkride oral exam will usually constitute no more than about 30-40% of your time with the examiner. The DPE is trying to be sure you took your weaknesses in each area seriously and that you and your instructor, as required, went back and remedied your lack of knowledge. But in the end, most of what the checkride is about is your ability to fly the aircraft effectively and safely. It's about proper radio calls, coordinated turns, safe landings, go-arounds, and a million other things you can only test in the air. Most of what the written test is about is ensuring that you have a good fix on the knowledge you need to fly in the airspace system.

There is overlap between the two types of assessment to be sure. In fact, the *Airman Certification Standards* introduction notes

---

[34] You could also have your checkride administered by an FAA Aviation Safety Inspector (ASI). But as an FAA paper on using an ASI states, "To take this route, however, you would need a fairly compelling reason. The wait can be long because resources at the local Flight Standards District Office [FSDO] are limited. Most people prefer to go with a DPE."

that "the *ACS* integrates the elements of knowledge, risk management, and skill listed in 14 CFR part 61 for each airman certificate or rating." I like to think of the three elements listed—knowledge, risk management, and skill—as part of a continuum from "book knowledge" to "flying skill." The "knowledge" element box can, by definition, be checked through your work to pass the written exam. The "risk management" part can be assessed to some degree in a written exam but will be probed during the oral and flight portion of your practical exam. The "skill" element is one that can only be assessed while you are at the controls of an aircraft.

One final argument Wildes makes is that the format of the exam is unfair, noting that "the rigid multiple-choice structure of [the exam] doesn't provide any wiggle room for students who might want to clarify something." Putting aside for the moment that delivering an exam in large numbers every year in more than 300 testing centers across the United States limits your options with respect to how a test can be effectively administered and quickly graded, flying is not a hobby or profession for "wiggle room." The correct answers to questions dealing with issues like what code you put into your transponder during an emergency, how you calculate weight and balance, and the rules for entering Class B airspace, are discreet. There is no nuance. Nor should there be. At best, "wiggle room" in the real world will get you in trouble; at worst, it will get you killed.

I sound like I'm picking on Mr. Wildes. That's not my intention. As I said, I'm sure many of you reading this would embrace some of his arguments. And I admit that I can't *prove* that the written exam as its currently structured makes a better and safer pilot. But my suspicion is that it does. I certainly don't think it does harm. Would you want to share airspace with people who didn't have the knowledge you do? I wouldn't. It's arduous to have to study in preparation for the exam and there is, no doubt, some stress induced when taking it. These things, to my mind though, are good. If nothing else, they ensure that only those serious about flying will make the cut.

Preparing for the written test

The preceding discussion is interesting if you like to debate the merits of something like a standardized test. But to some extent, it's all noise. Because, for now, the FAA *does* require you to pass a written exam. You might as well get ready for it.

But how? I noted earlier that the FAA has wonderful free resources that include all the information you need to pass the exam. There are a select few who probably can use just those publications to prepare and pass the knowledge test. But I wasn't one of them and the odds are you aren't, either. I therefore embarked on finding an online program that could help me get ready.

There are several very prominent online options to choose from. While they may vary in the particulars of how they present information, virtually all use video, text, animations, and quizzes to help you become practiced in critical concepts. The main differences you'll find are differences in price and access. Some offerings only allow you to access their content and learning tools for a defined period. Some allow access in perpetuity. I'd suggest choosing a plan that allows the ongoing access. I've found it helpful to refer to content every now and then when looking to refresh my memory about something. Also, as new concepts are introduced or new, clearer explanations for a concept are created, continued access allows you to benefit long after you've taken your written test.

I chose Sporty's as my provider and found their interface and content quite good. (I make a living in the online education space so have some experience in the area.) That said, I know others who have chosen competitors to Sporty's and have been quite happy. All the courses do something that you will probably find hard to do on your own: organize information, hold yourself accountable through quizzes, and focus on areas that are most important and therefore most likely to be included in the written test. This latter point is a crucial one. The universe of things, per the *ACS* standards, that could be asked of you on the written test is enormous. Most online ground school companies are well-versed in the categories of questions most likely to be asked on your written exam. This allows them to focus and structure their content in an efficient manner that helps ensure your success. To be clear, I'm not suggesting that this is "gaming" the system. The areas covered have been determined to be the most important. Might as well use others' work to focus on those areas.

While online ground school products do a good job of providing you with a structured syllabus, you will still have to "do the work." Most people won't be able to just watch the content and take a quiz before moving on to the next concept and then have confidence they'll recall information when test time comes. Even for those who can, there's a risk that the knowledge is superficial—that it doesn't get to the heart of "why," but instead just provides some vague reminder that makes you select one answer over another. This, in fact, is what I set out *not* to do. I did my best to understand underlying concepts. Doing so makes it much more likely you'll do well on the exam and, more importantly, will make you a better pilot.

The process by which you'll get that level deeper—i.e., get to the "why"—will be unique to you. I took copious notes in a spiral notebook with a selection of multicolored highlighters at the ready. I'd stop my ground school videos frequently to make notes on the pages and would use the highlighters to accentuate important points, using different colors to try to categorize my notes (orange, for example, was for weather-related concepts).

I also drew a lot of pictures and even downloaded images from the internet, cut them out, and pasted them into the notebook. Aviation can be very visual in nature—think of your instruments, sectional maps, checklists, and a traffic pattern diagram. All communicate information through shapes, colors, symbols, numbers, or the like. In fact, symbology, colors, and diagrams are as core to aviation as acronyms are to the military. It's not an accident that visuals prevail. Our brains are wired to recall images much better than text. Called the *picture superiority effect*, this phenomenon is well studied. Your ability to recall images is about six times greater several days after seeing them than your ability to recall text correlated to the images. And here's another cool thing: if you draw your own pictures, you also help recall. In an experiment conducted at the University of Waterloo, some subjects were told to write down words they were told, and some were told to draw pictures of the words. Those who drew pictures recalled the words at twice the rate of who just wrote them down.

Invariably, there will be a concept you will have trouble grasping. When that happens, the internet will be your friend. A moment ago, I did a Google search using the phrase "airplane in traffic pattern." It yielded more than 200,000 video returns and nearly 12 million total returns. Some of the results undoubtedly are not giving me what I want. But those at the top of the list of results certainly were. Consuming varied content like this can be helpful.

YouTuviators

To that last point, if you take your education seriously, then you will invariably begin watching YouTube personalities who offer instruction, analyze crashes, and show themselves having fun. I encourage viewing these. In fact, I'd watch everything aviation-related you can. That said, there are some folks making videos who shouldn't be. Some just give bad advice or "instruction." Some get personal when they don't need to, using their YouTube channel to settle scores or bash competitors. Some are just boring. Just have your antennae up when watching these so you can separate the wheat from the chaff.

Because there are content producers who share videos of themselves or others doing stupid—and sometimes illegal—things and getting away with it, some in our community caution against watching any of it, worried that impressionable pilots will mimic the behavior. This is a valid concern. But on balance, I think there's more to learn by casting your net wide in the world of pilot-influencers than not. If you are serious about your studies and flying safely, you will be able to recognize advice or actions that are sketchy. And if you are the sort of person who is anti-authority or likes to take risks, then watching only the "good" videos won't matter anyway; you will do what you want to do.

There are three broad categories of these sorts of video bloggers (a.k.a., vloggers): instructional vloggers, accident lessons learned vloggers, and entertainment-only vloggers. I used the instructional vloggers frequently during my training. It was particularly helpful to watch many videos about a topic that was harder for me to grasp. Explanations always varied slightly and often came with graphics, animations, or actual in-flight video to illustrate a concept. This was helpful when it came to internalizing information for later recall. More importantly, seeking out many explanations for the same thing helped me understand underlying concepts. This is really what you are looking to do. As I noted earlier, while some things are just memorization—time frame to report an accident to the FAA for example—you will find the written test a lot easier if you get the "why" behind something. For example, understanding that an aircraft's takeoff roll on a warm day will be longer than on a cold day is good to know. But grasping the "why" behind it will also help you understand better how a carbureted engine performs in different conditions. This, in turn, helps you understand why you can and should lean an engine in some scenarios. Having underlying knowledge allows you to pull at a thread that will give you a sort of unifying theory of how seemingly unrelated things are connected. And that, my friend, won't only make the written exam easier, but will also make you a much better pilot.

The second type of video content I consume religiously is that dealing with aircraft accidents. Videos that provide detailed retrospectives of what led to an accident are very instructive. *Every* pilot should watch these. If I were in charge, I'd yank the certificates of those who didn't. As I mentioned earlier in this book, almost all accidents are the result of pilot error. And almost all pilot error comes down to poor aeronautical decision making (ADM). Watching breakdowns of accidents is a way to remind yourself of the dangers of letting good judgment lapse. I highly recommend the videos put together by the AOPA's Air Safety Institute. They use flight simulator graphics to re-create accidents, often with real audio, in a way that drives home important lessons.

Many of these videos are usually of accidents that happened well in the past. A slew of other aviation YouTube channels cover more recent accidents, sometimes the very next day. These too are very helpful. Often, the YouTuviators who create this sort of content are making guesses about the cause of a crash—they can do little else given the recency of the incidents they are covering. I don't get too bent out of shape about that although I understand why others might. My reason for watching and encouraging others to do so is simple: the vloggers are almost always right. And do you know why? Because pilots keep killing themselves in the same ways over and over again. More than anything, this is what you learn by watching content like this. I find watching these videos the best way, *by far*, to keep myself vigilant about my practices when I'm in the cockpit. Note that these videos can be disturbing to watch. But making yourself disturbed is a small price to pay if the alternative is making yourself dead. Watch these and other videos. Watch every single one you can get your hands on.

Finally, there are those vloggers who just aim to entertain, although often while educating as well. These helped a *ton* to keep me motivated while in training. As I noted earlier, training can, at times, be a grind. It's easy to forget why you are getting your PPL. You are *not* getting it to do steep turns or stalls straight ahead. You are presumably getting it to take the family to visit relatives, or conduct business, or view fall colors from a different perspective, or get that very expensive hamburger. If you can't do those things yourself while training (and most can't), then the next best thing is to watch others do it. AOPA estimates that 80% of those who start flight lessons never get their certificate. Undoubtedly some of the reason for that is an inability for a student pilot to "keep their eye on the prize" by imagining the freedom that will come once the certificate is firmly in hand, allowing them to hop in a plane and go where they want. So, live vicariously through others and start thinking about what *you* will do when it is totally up to you where to go and what to do.

Interlude: Fun after your PPL is in hand

A quick aside: even watching others on YouTube regularly didn't prepare me for the fun and flexibility I ultimately did realize after getting my PPL. One thing nice about living in New England is that a lot of very cool destinations are within a couple hours of flight of any departure point. A flight my wife and I have come to enjoy is one to Martha's Vineyard, an island off the south side of Cape Cod. Known mainly as a summer playground for the wealthy who live on the East Coast somewhere between New York and Boston, it's a small island (though about twice as big as its famous sister, Nantucket Island) with two larger enclaves comprised mainly of shops, restaurants, and inns. The beaches are a big draw. Because it's an island, most visitors get there by ferry. Not us. What would be an impossible day trip in the summertime—or anytime for that matter—is a breeze with a general aviation aircraft and a private certificate. During the summer months, the trip from our house to the Vineyard by car could be anywhere from three to five hours. In an airplane? Forty minutes. We can leave in the morning and be back at our home before dinner. Watching YouTuviators gave me a taste of what it would be like to take a trip like this before I was capable of doing it myself. And while the reality was even better than the dream of it, watching others' adventures was a nice reprieve from studying and repeating maneuvers unrelentingly.

My written test experience

I took my written exam just two months after my discovery flight and after passing two practice exams on the Sporty's platform. I needed to pass the two tests to get an endorsement from Sporty's to take the actual FAA exam. You'll hear the term "endorsement" thrown around a lot in aviation in relation to training. The definition of "endorsement" according to the *Oxford Dictionary* is "an act of giving one's public approval or support to someone" and that's exactly what it means in an aviation context. Some authority—usually a CFI—must give you an endorsement before you are allowed to embark on a new-to-you task. The written exam is the one activity where the endorsement probably won't come from an instructor, at least not a human one. With the advent of the internet, online schools began replacing instruction by humans (at least in a Part 61 school). If you use an online ground school, your endorsement will come from a computer that believes you are ready to take the written exam based on your performance on practice tests.

The flight school I attended had a testing center at Hanscom. It was in a small room adjacent to the main counter where pilots and students came and went all day long. It had an entire wall of glass, in fact, that looked into the main area. This concerned me. I was worried about distractions related to ringing phones, people checking out aircraft, and instructors talking to students. The school had anticipated this and provided ear plugs. I was also provided some pencils, paper, a straightedge, and the FAA booklet that contained figures the test would reference. This booklet is available from the FAA's website. I recommend downloading it and getting to know it well. While it's just full of images with no qucstions, you will be asked to reference those figures during the exam. Having a sense of what the images are and the possible questions you could get related to them helps get your brain prepared to use them.

As I was getting situated to take the exam, a new employee was trying to set up the computer I would use. And wouldn't you know it? She was having trouble connecting to the testing service. I watched with some trepidation as she fumbled with a mouse and then powered the computer on and off. I had paced my studying to reach a crescendo a couple of days before the exam and didn't relish the idea of having to cancel the effort that day and then return another. I had a form of "get-there-itis"—I wanted to go *now*, not later. Eventually, she sought help from another employee who had more experience and seemed to know what to do to get the system to work. I breathed a sigh of relief. Remember when I said that part of the reason the exam is valuable is that it can, at a basic level, ensure you are able to handle some level of pressure? While a finicky computer is not the same as a lost engine, it drove the point home. Can you stay cool and collected when things aren't going your way?

I had heard that the major online schools did a good job of preparing you for the exam, that there would be few surprises. And that turned out to be the case for me. Many of the questions I had seen before—not necessarily with the exact same numbers or scenarios used, but certainly with similar structure. There was only one question I felt uncertain about. It had to do with speed limits inside of an airspace. I hadn't paid too much attention to these limitations. I did remember the numbers 200 knots and 250 knots as the only options for what the speed limit could be but couldn't recall the details of when those limits were in effect. Admittedly, part of the reason I hadn't paid enough attention to this during my studying was that the planes I was flying couldn't get to those speeds without being ripped apart. I made my best guess and moved on.

I finished well within the time allotted to take the exam. If you used all 120 minutes to take the exam, then you would have an average of about one minute and 50 seconds for each of the 65 questions. That may not seem like much but remember many questions you'll answer in 15 seconds or less. So, you'll have plenty of time on the questions that stump you or require some calculations.

After I reviewed my answers, I clicked the "submit" button and stepped outside of the testing room to let the administrator know I was done. She came back in and printed the result.

"You did great," she said as she looked at a printout sliding out of a printer in the corner.

"How well?" I asked, smiling.

"Ninety-seven percent," she said with a flourish as she waved the paper in the air. I missed two questions. I was relieved.

When she handed me the printout, I looked to the section right below the header information where the following was written:

> *The Airman Certification Standards (ACS) codes listed below represent incorrectly answered questions … A single code may represent more than one incorrect response.*

Written below this statement were the codes PA.I.E.K1 and PA.III.B.K1. These correspond, as the helpful message above notes, to *ACS* requirements. The FAA doesn't tell you the questions you missed. They just note the subject areas the missed questions fell into. You are therefore left with only a directional idea of where you were wrong. This isn't so bad if you miss a couple of questions, but if you miss a larger number, you could have several codes listed that encompass wide swaths of the body of knowledge you're expected to master, and this can make it difficult to pinpoint where you should put extra study effort. It also makes you more exposed during the oral portion of your checkride since the DPE will focus on areas you missed during your written exam.

In my case, the two knowledge areas the codes represented were "types of airspace/airspace classes and associated requirements and limitations" and "towered and nontowered airport operations." I'm certain the first category was where the missed question regarding airspeed limits fell. As for the second category, to this day I don't know what question I answered incorrectly. But in the end, my relief at having successfully checked off one more item on my path to a PPL overtook my obsession about a missed question. I stepped out into the cool April air with a bounce in my step.

Chapter 11
Some Important Aerodynamics

My training continued into the spring months. I became *very* comfortable with high, gusty winds, a common feature of weather in Massachusetts during that time of year. In fact, and surprising to most (even those who live there), Boston is ranked anywhere between the seventh and third windiest city in the U.S. depending on what data you're looking at. New England sits under a spot where major weather systems converge, and residents feel it in the winds of spring and the Nor'easters of the winter. Just 130 miles north of Boston, Mt. Washington's 6,288-foot peak has seen winds over 200 miles per hour; it holds the record for the highest non-tornadic winds recorded in the world. Sometimes it felt like I was flying in those Mt. Washington winds.

"What do you think?" I asked Jeff after we had completed an engine runup one day and I switched to tower frequency on our radio in anticipation of takeoff. The tower had just issued a low-level wind shear alert based on the report of a landing pilot who noted he had seen wind speeds increase and decrease by 10 knots while on final approach.

"We'll be okay," Jeff said.

I admit to being nervous about this. "Low level wind shear warning," even to the uninitiated, sounds ominous. And it is.

The FAA definition of wind shear is "a change in wind speed and/or direction over a short distance." Wind is almost always variable in velocity or direction to some extent—if you've ever paid attention to a windsock for more than 10 seconds, you have probably seen it stretch and then go slack (indicating a change in wind speed) or pivot about the pole it's attached to (indicating a change in wind direction). But severe and rapid changes in wind speed or direction can be deadly.

Recall that the thing that makes a wing produce lift is airflow over it, and that the wing doesn't know or care what the cause of that airflow is—it could be because the engine is driving the wing through the air, or it could be because a wind is blowing over the wing. Or, and this is almost always the case, it can be because of both.

During landing, you are attempting to slow the aircraft to stall speed right as the wheels touch or are just barely above the ground. This ensures the slowest touchdown speed. Slow touchdown speeds are safest for a plane; they are easier on the airframe and ensure shorter landing distances. To help you get your ground speed as slow as possible, you are, as much as practical, executing your landing into the wind. Therefore, a component of nature's wind (in addition to the wind created by moving through the air) is being registered in your pitot (pronounced PEE-toh) tube, the device on the outside of the plane that senses (indirectly) how fast air is moving over the wing.[35]

---

[35] If the pronunciation of "pitot" sounds distinctly French, that's because the word comes from the name of the Frenchman who invented the pitot tube, hydraulic engineer Henri Pitot. It's interesting to note that Bernoulli and Pitot were contemporaries, and they did their research with water, not air. But

The pitot tube measures something called "ram air." This might be one of the most self-descriptive of aviation technical terms, because, at its most basic level, the pitot tube senses how much air is being rammed into it. But there's a bit of nuance to this. The pitot isn't really measuring the speed of the air entering it, it's measuring the *pressure* created by the air moving into it. This pressure is then compared to static air pressure. Static air pressure is the pressure of the ambient air outside the aircraft; it is the pressure caused by air that is not moving. The difference between ram air pressure and static air pressure is then translated into airspeed; this is what you see on your airspeed indicator.

Keep in mind that the pitot tube doesn't know or care if the airplane is actually moving. If you had a sensitive enough airspeed indication system, your airplane's airspeed indicator would register 10 knots even if the plane was stationary on the tarmac but pointed into a 10-knot wind. In fact, to help us dive into the concept of airspeed further, let's imagine you *did* have a super-sensitive airspeed indicator and it indicated 10 knots of airspeed when a 10-knot wind passed through the pitot tube. If you needed 60 knots of airspeed to get airborne, then you just need an additional 50 knots of moving air since you already have 10 knots due to the wind. That's what your engine is for: to generate that additional wind to ram through your pitot tube. Well, that's a secondary purpose. Of course, the main purpose is to get air over the *wing* in sufficient quantity to produce lift.

Effectively, whenever you have a headwind component while flying, the engine is "filling the gap" between the airspeed provided by the headwind and the airspeed required to fly. The

---

as I noted earlier, air is a fluid like water and thus their findings were useful about 150 years later when aviation came on the scene.

larger the headwind, the smaller the gap the engine must fill. The launch of B-25 bombers from a carrier deck during the "Doolittle Raid" of *Thirty Seconds Over Tokyo* fame early in World War II illustrates this point well. Launching the planes required that gap between the oncoming wind and the takeoff speed of the bombers to be very small. With little real estate to work with and heavy bombers not designed for carrier operations, a headwind wasn't just desired, it was necessary. Fortunately for the crews, they had one. According to Lt. Col. Jimmy Doolittle himself, when one pilot got into his B-25s, "the plane's airspeed indicator showed about 45 miles per hour sitting there.... This meant he had to accelerate only about 23 miles per hour" to get airborne. The wing didn't know that most of the airspeed was coming from the forward motion of the ship through a headwind. And it didn't care.

With that background, let's get back to our discussion of wind shear, which is the flip side of the headwind-helps-you-takeoff coin. Assume you are on final approach with a 10-knot headwind and your airspeed indicator is registering 60 knots. If, suddenly, that headwind dies, you will have only 50 knots of airspeed. That's a problem if your stall speed is 53 knots. Not enough air is flowing over the wing to maintain lift. Your stall warning horn will be screeching like an injured cat.

The danger of changes in wind speed, even if a wind shear alert is not in effect, is why, when winds are very gusty, you should add a "gust factor" to your approach speed. The generally accepted math when it comes to adding to your approach speed on a gusty day is to add half of the gust factor to your final approach speed. For example, if the wind is 10 knots gusting to 18, you have a gust factor of 8 knots (18-10). Half of this is 4 knots. Add this to your normal approach speed; if that is 65 knots, make your adjusted speed 69 knots (65+4).

One reason Jeff wasn't that concerned about the low-level wind shear warning is that while wind shear isn't something to be taken lightly by any pilot, it's something that's more dangerous to aircraft powered by turbines (i.e., jets). This is because turbines don't react as quickly to pilot inputs for increases in power; they take time to "spool up." Therefore, the pilot of a jet aircraft must anticipate the changes they will need in power. Since wind shear can be sudden and dramatic, this puts the pilot at a bit more risk if they find the wind that was blowing in their face one minute is gone the next and they need extra engine output to increase airspeed to remain clear of a stall. A piston engine, on the other hand, reacts almost immediately, just as a car engine does. This means that the attentive pilot of a piston-driven, light, general aviation aircraft can compensate more quickly for a shear-induced increase or decrease in airspeed.

Jeff and I took off and, while landings were a challenge, I indeed didn't find the experience nearly as traumatic as the nefarious term "wind shear" would indicate. That said, I forced myself not to be complacent. Smooth air of a sufficient velocity over the wing is all that matters and if conditions exist that can decay the velocity or "unsmooth" the air, you could end up with some big problems.

While I longed for a dead calm day, all that training in windy conditions was paying off. I was getting good at landing with gusts and crosswinds buffeting the aircraft. If there is one skill that I believe is critical to master if you hope to really use your PPL to fly regularly, it is the crosswind landing. While runway orientations are usually chosen based on the direction of prevailing winds in the area around the airport to increase the chances the wind is down the runway, Mother Nature isn't as accommodating as we'd like. I can think of only a handful of times that the wind was blowing consistently and precisely out of 290° when I was landing on Runway 29.[36]

Slip sliding away ...

One of the difficult things about learning how to properly execute a crosswind landing is that it requires you to do something I just spent a lot of time saying you *shouldn't* do: fly in an uncoordinated fashion.

One day, when landing with a left crosswind, Jeff told me to drop the left wing by turning the yoke to the left while pushing on the right rudder pedal. This is "cross-controlling" the airplane; you are inputting rudder in the opposite direction of your roll. Your ball will be way off center in your turn coordinator.

"Wow, that's uncomfortable," I said to him when I timidly did as he asked. He pushed me to be more aggressive with both the dip of the wing and the use of the rudder.

"I feel like I'm rubbing my stomach while I pat my head," I said. The principle of entering rudder in the direction I turned the yoke was so ingrained that asking me to do the opposite was like asking me to write with my non-dominant hand.

---

[36] One advantage early aviators had over us is that there were no runways; they took off from and landed in large, open fields. This meant they could always take off and land into the wind. Orville and Wilbur had it easy!

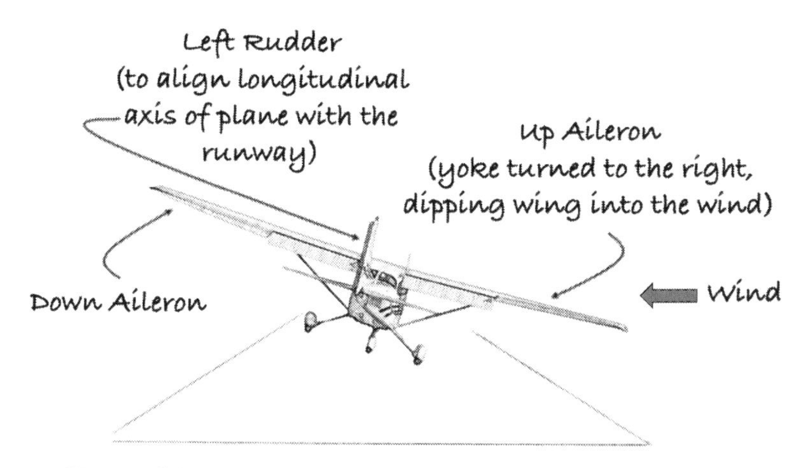

You will likely have the same reaction. Don't fret if you do. It takes time and, eventually, becomes second nature. The dipping of the wing into the wind keeps the airplane from drifting off the centerline of the runway. The problem is that this will cause the nose of the airplane to point somewhere other than straight down the runway. The rudder input in the opposite direction is to swing the nose back around to align the longitudinal axis of the plane with the centerline of the runway.

This of course begs the question of why this is okay if coordinated flight in every other circumstance is considered sacrosanct? It's okay because it's one of the best ways to execute a safe crosswind landing. But it's only okay if done properly. We talked about the dangers of uncoordinated flight earlier in a very specific context: when entering a stall. This is because a stall when combined with yaw (i.e., you're uncoordinated) is what leads to a loss of control in the form of a spin. If you execute a crosswind landing correctly, you have some yaw due to your rudder input, but you shouldn't have a stall. You are leaving out one of the ingredients that leads to a spin. The key to safety when cross-controlling for the purpose of landing in a crosswind is to keep your airspeed in an allowable range by keeping the nose down. This decreases the chances of a stall. Being slow with a high angle of attack—with or without a lot of power—is the dangerous place to be. So, watch your speed and keep the nose low ... you're trying to descend after all.

Also note that what I'm describing for a crosswind landing is a slip, not a skid. In a skid, the airplane is also in an uncoordinated state, but the skid is dangerous; the slip, when done correctly, isn't. The most common reason you might (inadvertently) put an airplane into a skid is while trying to "pull the nose around" after you have overshot the centerline of the runway on a base-to-final turn—you are, for example, making a left turn to final only to find that you are well right of the centerline when you begin to roll out. Rather than turning back to the centerline in a shallow, coordinated turn or (and this is *always* an option) doing a go-around, you push on the left rudder pedal to force the nose back toward the runway while banking even more aggressively. This causes the ball in your turn coordinator to get uncentered; it will be offset to the right. But you may not notice. After all, you are applying rudder in the direction of the turn just as you do to overcome adverse yaw in any run-of-the-mill turn. That is the insidious thing about a skid. You are getting uncoordinated without having to cross-control the aircraft.

If you did look at your turn coordinator in this scenario, it would be obvious that you need to release pressure on the left rudder pedal to bring the ball back to center. If you don't, your left rudder input will make the plane want to turn into the low wing even more and the high wing (the right one in our example) will come up. While it comes up, the high wing will also begin to sweep around the outside of the turn, moving faster than the low wing and therefore generating more lift. This will cause it to raise even more, making the turn to the left become even more pronounced. This probably sounds familiar; this is the beginning of a sequence of events that can result in a spin as described earlier. To correct for this increasing rate of turn to the left, you will probably (unconsciously) turn the yoke to the right, trying to arrest the continued roll to the left. Makes sense, right?

But this is where things go from bad to worse. Turning the yoke to the right to try and correct for the roll to the left will raise the right aileron and lower the left. This has the effect of making the angle of attack on the low (left) wing *higher*. Why?

Recall that the angle of attack is defined as the angle between the relative wind and the wing chord, a line drawn from the leading edge of the wing to the trailing edge. Therefore, there are only two ways to change the AOA: you can change the angle at which the relative wind hits the wing, or you can *change the angle of the wing chord relative to the wind.* A deflected aileron does the latter.[37]

When you move an aileron down, you change where the trailing edge of the wing is in space and therefore increase the angle at which the wing chord hits the relative wind. You have made the AOA on that wing greater. Increased AOA also increases drag, which wants to pull the left wing "back" just as we described in the discussion about adverse yaw, causing the plane to want to turn even more to the left. If the critical AOA is eventually exceeded, the left wing will suddenly stall, causing the airplane to "snap" into a roll to the left and throw you into a

---

[37] You may have already figured out that deployed flaps also change the angle the wing chord hits the relative wind, increasing AOA, lift, and, therefore, drag. Higher drag, which translates to slower speeds at a given power setting, and more lift while throttling down are good things when you're trying to land.

spin. Given that you are probably no more than 500 feet AGL at this point, recovery will not be possible.

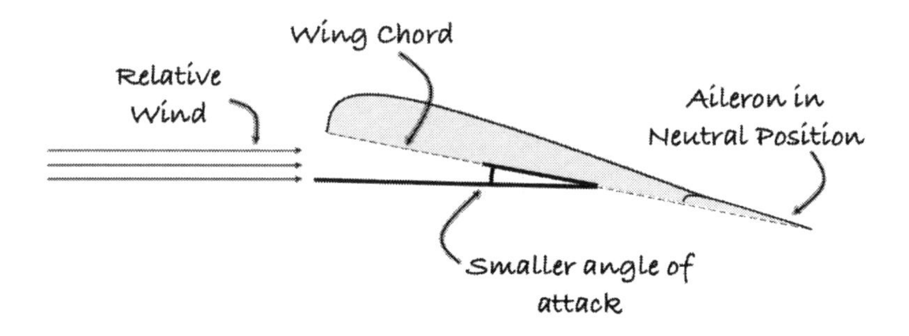

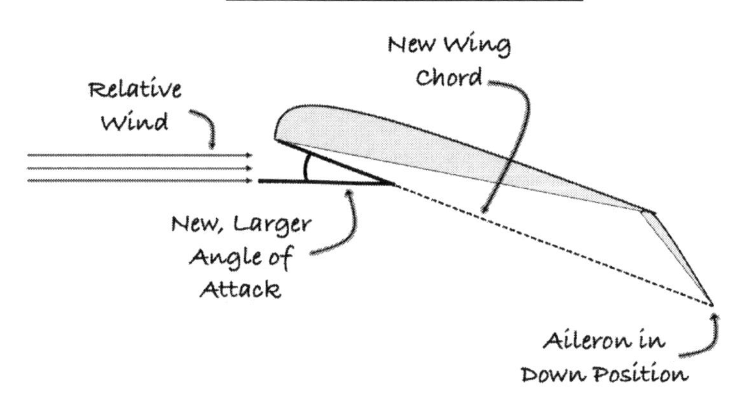

In addition to these factors, there are a few others that contribute to the tendency to enter a spin when in a skid. First, because the low wing (the left one in our example) is shielded from the oncoming airflow by the body of the airplane (since it is yawed the body of the plane "screens" the wing), the wing is robbed of the lifeblood of lift—smooth air moving over it. Finally, you will likely want to aid your turn back to the centerline by pulling back on the yoke, which increases the load on the wing, making the stall speed higher.

Yes, there are a *lot* of factors working against you in a skid that goes too far. If you want to remember *not* to do this while turning base to final, feel free to recite my little ditty below while turning back toward the runway:

When you overshoot final and need to get back,
Be careful with the rudder or you might go splat!

As I noted earlier, a slip, our other uncoordinated state that we use for crosswind landings, is not dangerous when executed properly. And, in fact, there are some elements of a slip that make it self-correcting when it comes to stalling a wing. Imagine you have a particularly strong wind from the left, and you must keep lowering the left wing, and therefore raising the right wing, to counteract it. Recall that this will deflect the right aileron down and therefore increase the lift, AOA, and drag on that wing. In contrast to a skid, though, you'll be entering rudder input to the right, in the opposite direction of the turn. This means you aren't helping to sweep the high wing around the nose of the plane, which is what happens when a skid turns into a stall/spin. Instead, if, while trying to get aligned with the centerline of the runway, you happen to exceed the critical AOA on that high wing, it will fall in a stall. But because it's the high wing, it will drop, and this will have the effect of returning the airplane to a more coordinated state. *But*—and this is a big "but"—you are still in uncoordinated, stalled flight. Those are the two conditions you need to create a spin. So, while a slip is less dangerous than a skid, you should never exceed the critical angle of attack on either wing. But hopefully, this explanation helps you see why a slip isn't as worrisome as a skid.

There's another aspect of a slip that is different than a skid. In a slip, if the fuselage of the airplane is going to be shielding a wing from oncoming air, it will be the *high* wing. If you must choose between having the low wing shielded or the high wing, it's the high wing you'd pick. Again, this is because if that wing stalls, it will force the airplane into a more coordinated state.

Side or forward?

We've called this crosswind landing maneuver a slip. And while that's correct, it's more correct to call it a *side* slip. This is to distinguish it from a *forward* slip. What's the difference? From an aerodynamic perspective, nothing. Both are cross-controlled maneuvers. But in a side slip while landing, you are putting in just enough rudder to line up the nose of the airplane with the runway centerline. This has the effect of pointing the airplane in the direction it happens to be traveling; your nose is aligned with your ground track. And, hopefully, that ground track is aligned with the centerline of the runway since you intend to land there.

In a forward slip, you are typically inputting a lot more rudder in the opposite direction of your aileron input than in the case of a side slip. In this maneuver you're not concerned with getting the nose pointed down a runway centerline. Instead, you are trying to lose altitude without a significant increase in airspeed.[38] A forward slip helps you do this because, since your nose is not pointed in the direction of flight, the fuselage of the airplane

[38] I find the naming of the two versions of slips very confusing. If you do too, remember that you use a side slip when you have wind from the side on landing (i.e., you are correcting for a crosswind).

becomes exposed to the oncoming air, creating a lot of drag. This drag has the effect of slowing the aircraft, causing a loss of lift and subsequent loss of altitude. That said, it's important in both a forward and side slip to keep the nose of the airplane below the horizon. You don't want to risk becoming too slow and entering a stall. Also, because your pitot tube will not be pointing directly into the oncoming air, your airspeed indicator could be erratic and inaccurate. (Note that some aircraft POHs advice against slips in certain scenarios; be sure to check if such restrictions apply to your aircraft.)

Forward slips are very commonly used in aircraft without flaps. Since flaps help provide additional lift at slower speeds during landing, not having them means a pilot is lacking a tool to slow down and descend. The forward slip can therefore serve as that pilot's "virtual flaps." But even if the aircraft you fly does have flaps (which is likely true for 99% of readers), you should still know how to do a forward slip in the event flaps malfunction and won't deploy or you must execute a forced landing after the loss of engine power and don't have the ability to "fine tune" your approach.

To that last point, imagine a scenario where you've lost your engine but, fortunately, you identify a beautifully mowed large field nearby that looks perfect for use as an emergency landing location. The field is big, but trees line it on all sides. Unfortunately, as you approach your improvised landing strip, you realize that you are way too high; your efforts to conserve energy and ensure you make it to the field have worked *too* well. Just pushing forward on the yoke or stick and pointing the nose at the field to try and get down is going to increase your airspeed and you realize that doing this might carry you through the field into the trees on the far end of the open area. This is where a forward slip comes in handy. It can help you descend rapidly *without* that commensurate increase in airspeed. Also, in a case where the field is short and you must "drop in" just beyond the trees at the near end of your landing spot, the forward slip can help you do that too.

If you are questioning the value of a forward slip, consider the story of Air Canada Flight 143. As is the case in most aviation incidents and accidents, a long litany of unfortunate events led to the unusual story of Flight 143. I won't go into all of those here but suffice it to say the Boeing 767, a wide-body, transcontinental, and transoceanic aircraft that can hold roughly 216 passengers (although only 61 were on Flight 143) ran out of gas while cruising at 41,000 feet on a flight from Montreal to Edmonton in 1984. The crew reported that when they looked to their emergency procedures to determine how they were going to glide the jet to a safe landing, there was no "both engines out" section. Apparently, Boeing engineers lacked the creativity to consider that this might ever be a scenario a crew would face.

Because it was never determined and published by Boeing, the pilot, Bob Pearson, made a guess at V_g, best glide speed (in the Warrior it is 73 knots; Pearson guessed that the 767's was 220 knots). This is the speed at which the aircraft will lose the least altitude for a given distance over the ground. It helps pilots without engines stay aloft as long as possible, increasing options when it comes to finding an emergency landing spot or trying to remedy an anomaly. While maintaining 220 knots, Pearson's first officer, Maurice Quintal, did some calculations that, with ATCs help, provided estimates of the amount of ground they had covered and altitude they'd lost during a defined stretch of time.[39] These calculations made obvious that an emergency landing at Winnipeg would be impossible; they'd run out of altitude before getting there. Fortunately, Quintal had served as a pilot in the Royal Canadian Air Force and realized an abandoned Air Force base he had once flown out of seemed within range of their recently acquired glider. The name of the base was RCAF Station Gimli.

Quintal didn't know it, but Gimli had been turned into an auto racing complex with the old runways serving as drag strips. A metal guardrail had been installed down the center of one runway surface to divide it into two racing lanes. While he and Pearson were planning their landing at Gimli, the Winnipeg Sports Car Club was hosting a "family day" event there. The area was covered with families enjoying the too-infrequent summer days Canada experiences. They probably didn't expect that the emergency landing of a wide-body jet would become part of the program.

With both engines out, the 767 lost the energy used to power some hydraulic systems, so Pearson and Quintal found themselves without flaps and slats, both used to help slow the aircraft in preparation for landing. When Gimli came into sight, they realized they were too high and too fast to land, but not high and fast enough to circle around the field one more time before

---

[39] This information allowed Quintal to estimate that their glide ratio was about 12:1, meaning that for every 12,000 feet (approximately two nautical miles) they moved forward, they lost 1,000 feet in altitude. I found this ratio incredibly hard to believe since a Piper Warrior also has a glide ratio of about 12:1. A separate source said that the glide ratio of the airliner is closer to *20:1*. I don't know if this means the 767 is a great glider or the Warrior is a horrible one.

putting the aircraft down. In one of the most daring examples of "what do I have to lose?" thinking, Pearson, an experienced glider pilot, decided to use a forward slip to lose altitude and keep his airspeed in check. He dipped the left wing significantly and jammed on the right rudder. In an article in *Soaring Magazine* a passenger said that the unusual maneuver had the effect of making him look straight down at a golf course so close to the ground he could almost see what clubs a group of golfers was using.

Pearson's maneuver worked. He landed the aircraft with his main landing gear straddling the guard rail in the center of the runway. This was fortuitous. Because the nose gear never locked in place, it collapsed upon touchdown so lateral control of the aircraft was lost when the nosewheel steering was lost. The guardrail acted like one of the rails toy cars at amusement parks ride along, giving youngsters the illusion they are in control. The 767 was "steered" by the rail, keeping it from veering off into the crowds around the edges of the runway-turned-racetrack-turned-runway-again. Everybody on the airplane survived, and nobody on the ground was hurt or killed.

The story of the "Gimli Glider" shows in a very stark way the utility of a forward slip. It also shows that aerodynamics are aerodynamics. While aircraft of different sizes and designs have their own idiosyncrasies, the laws of physics apply across airframes.

Chapter 12
Solo Time

I progressed through my training at an average pace. I would have loved to have been going faster, but the fits and starts I alluded to earlier (weather, work schedule, etc.) kept getting in the way of making rapid progress. By the end of May, I was itching to solo. That's certainly not uncommon. It's a major milestone of private pilot training in a training regime that only has four: the written exam, solo, cross-country solo, and successful checkride completion.

Finally, after a lesson one day, Jeff said, "I think you'll be ready to solo the next lesson." As much as I had longed to hear those words, I suddenly and unexpectedly felt trepidation.

Was he insane?

He wanted to trust me to, all by myself, take an aircraft from the tarmac to the runway and into the air? And to bring it back down safely again? To talk to ground and tower without him ready to correct or guide? Where did he get his CFI certification? Through a $50 correspondence course?

Of course, he didn't know this was what I was thinking. I shrugged his suggestion off, matter-of-factly.

"Well, if you think I'm ready. . . ."

I returned home that evening and excitedly told my wife about the prospect of a solo. Her reaction didn't help my own doubts.

"Already?" she said, as if I had told her I was going on a mission to the moon after 19 hours of flight time in a Warrior.

"Yep," I said, again trying to sound casual. It was only one word, by I tried to infuse it with an I'm-not-concerned tone that I imagined mimicked the way Chuck Yeager might have responded to the question, "You going to break the sound barrier today?"

A few days later, I was driving back to Hanscom for what I thought would be my solo. During the drive, I spoke to myself as I imagined my takeoff roll.

"Engine instruments are in the green. RPMs are stable. Airspeed is alive. Forty knots. Forty-five. Fifty. Fifty-five ... rotate. Pitch for V_y." I was nervous.

Unfortunately, all that nervous energy was wasted. When I got to Hanscom, high winds that I hoped would have abated by the time I arrived at the airport were still stubbornly blowing. Apparently, they were unaware they were getting in the way of the world's next great aviator flying solo for the first time.

Jeff and I did some ground instruction instead. I went home disappointed.

May turned into June. I was on my way back to Hanscom on a relatively calm day. I didn't allow myself to get too high; I didn't even talk to myself as I drove to the airport as I had the first time I thought I might solo. One thing I did do though—something I did on every drive to the airport—was look at the sky. It was a typical late-spring sky (yes, early June is late spring in the Boston area); some cumulus clouds hung under a developing overcast layer, their puffy form contrasting in a dramatic way with the slate gray overhanging them. It was a pretty sky but an unsettled one. Showers were popping up in the distance.

Jeff and I got into N238ND and headed to the active runway. I completed three landings in the pattern. During each one of them Jeff was critiquing me, more than usual it seemed.

After the third landing and taxiing back to the school's parking area he said, "How are you feeling? You feel comfortable doing a solo flight?"

I responded, half kidding, "Do *you* feel comfortable with me doing a solo flight?"

He replied, "I'm wondering how *you* feel?"

This verbal tennis match was not giving me much confidence.

His nervousness made more sense to me in retrospect. Experienced pilots—as we've established in these pages a few times—make mistakes that get them killed and/or damage property and planes. Turning loose somebody with less than 20 hours of flight time must be a nerve-wracking moment for any CFI. Only Jeff had primary responsibility for my training. There was no cadre of teaching staff that shared in the responsibility to get me enough education and skill to fly alone (although I did have a progress check of about an hour with another CFI). While it wasn't obvious to me at the time, if you were in Jeff's shoes, you'd be nervous too. My performance, fair or not, would be a reflection on him and him alone.

"I think you're ready," Jeff said, finally breaking our back-and-forth.

Jeff crawled out of the cockpit and onto the wing.

"I'll be listening in!" he said, raising a handheld radio for me to see.

I gave him a thumbs up as he pushed the door closed. I reached over to lock the top and bottom latches.

After getting the engine restarted, I listened to the ATIS again and then checked my radios to be sure I had the right frequencies tuned—the ground frequency in the active box and tower in standby.

"Hanscom Ground, November 238 November Delta, student pilot solo, west ramp with Foxtrot, ready to taxi," I transmitted. Jeff had encouraged me to use the "student pilot solo" whenever introducing myself to a controller for the first time until I got my private pilot certificate. The idea was that controllers would know they had a novice on their hands and speak more slowly and, hopefully, pay a bit more attention to you. I recommend strongly you introduce yourself in the same way while doing your solo training.

"November 238 November Delta, Hanscom Ground," came the reply. The voice was that of a woman whom I had become familiar with over my months of training. If I could have picked who I first wanted to talk to at a moment such as that one, it would have been her. She had a voice that was at once soothing but authoritative. I considered her one of the better controllers among a bevy of them at Hanscom.

"Taxi to Runway 29 via Juliet, Echo," she continued.

As I was preparing to hit my mic button and confirm the instructions she had just given me, she unexpectedly spoke again.

"And 238 November Delta, be aware that there's an area of precipitation about five miles to the southwest of the field."

I had seen the showers in the distance on my last lap around the pattern with Jeff and hadn't thought much of them. They weren't related to any thunderstorm formation, and they seemed to be stationary. While there was no rule that would have prevented me from flying in the rain, it wouldn't be ideal for a first solo experience. Visibility would be reduced, and the runway would be slick. I found more nervousness creep between the joints of my hands, one gripping the throttle and the other the yoke with pressures that would have turned coal to diamonds. Why was she pointing out the weather?

"Yeah," I said, "thanks for the heads up. I'm just going to be staying in the pattern. If you could keep an eye on that for me, I'd appreciate it." As soon as I spoke asking for help, I wondered if I had been presumptuous. But after a short pause she said cheerily, "Will do."

I was relieved. Not just because she would keep an eye on the rain, but because it did seem that she indeed was giving me some special attention.

I began my taxi to Runway 29.

***

In the cargo compartment of the plane, I was carrying something that had been hanging on the wall of my study at home. It was a framed montage featuring a picture of the space shuttle *Discovery* lifting off on its maiden flight, two photographs from orbit looking outside the spacecraft, and a final one of my father, floating in the cabin on what was to be the first of his three missions into space. Below the photographs, there was a small American flag and a mission patch, both of which had been flown into space on the mission. And written beneath the pictures and between the flag and patch were the words: *Presented to Mom and Dad from your son Mike.* Dad's signature was scribbled across a line beneath the inscription. It had been Dad's gift to his parents after the mission and had been handed down to me after my grandparents passed away.

I had grabbed the frame on the way out the door at the last minute. My grandfather was a B-17 aircrew member in the Pacific during World War II. It was his early life in and around airplanes that had stoked my father's imagination when it came to all things space and aviation. My father's profession as an Air Force aviator and astronaut had, in turn, stoked my passion for aviation. It seemed appropriate to connect my first solo flight to that family history in some way. While my grandfather had died many years before and my father lived far from me, it was a small way to have them with me on the flight. I thought of them during my taxi to the active runway.

"Okay, looking left, looking right," I said to myself as I passed through an intersection where several taxiways met. "This is a hotspot, so be sure I'm on the right taxiway," I said. The FAA defines a hotspot as "a location on an airport movement area with a history or potential risk of collision or runway incursion, and where heightened attention by pilots and drivers is necessary." Hotspots appear on airport plates (i.e., maps) as brown circles with an "HS x" designation, where "x" is the number of the hotspot. The hotspot I was referencing during my taxi was the only one at Hanscom, so it was "HS 1."

Be sure to pay attention to hotspot designations whenever you fly, particularly when going to an unfamiliar airport. I have seen why HS 1 is designated as such at Hanscom. The location encompasses the intersection of five taxiways at angles to each other. More than once, I had seen aircraft in front of me turn onto the wrong taxiway because of the number of intersections at odd angles.

Once I'd made it to the runup area near the end of Runway 29 and completed my engine check, I verbalized what I'd do in the event of engine trouble on takeoff. By this point in my training, I think I'd read everything I could get my hands on about aviation and had watched hours of YouTube videos. From that effort, I had learned that while takeoff accidents were three times less common than landing accidents, they were 20 times more deadly. Losing an engine low and slow is a perfect setup for a stall-spin incident.

"When the engine has issues on the takeoff roll and before I'm airborne, I'm going to cut the throttle and get on the brakes," I said aloud. Jeff had taught me to use the language "*when* the engine has issues" instead of "*if* the engine has issues." The idea was to get your brain biased toward thinking that the engine would not work as expected; this would help you be more ready to address an issue. If things did work as expected, well … that would be the surprise, not the other way around.

"When the engine has issues after I'm airborne but with runway still available, I'm going to get the nose down, land straight ahead, and get on the brakes."

Also note I say "when the engine has issues," not "when the engine quits." This is because reduced power and not a complete loss of power is often the way an engine failure will manifest itself. Said another way, your engine has "failed" if it can't keep you airborne, even if it is still running. So be sure to be prepared for something other than a dramatic and complete engine shutdown.

"When the engine has issues after I'm airborne, with no runway left, but below 1,200 feet, I'm going to pitch for best glide—73 knots—and pick my best options in front of me, taking care not to bank the aircraft more than about thirty degrees." Where does the 1,200 feet come from? Hanscom's altitude above sea level is 132 feet. If my altimeter reads 1,200 feet, that means I'm about 1,000 feet AGL. In a light, single-engine, training aircraft, you'll need about 1,000 feet of altitude above the ground to make a turn back to the field. An engine failure below this altitude when you don't have any meaningful runway left below you is obviously the most fear-inducing. You don't have enough altitude to make it back around to land at the field (at least not without initiating a dangerous and radical maneuver—more on this in a moment).

To prepare for this scenario, become familiar with the terrain off the end of the runway bounded by a half-circle defined by a radius of about three miles and covering 180° of forward view. I knew, for example, that off the end of Runway 29 at Hanscom, turning left if I had problems was better than turning right. There were some marshes in that direction as well as a few plowed fields. None of them were great options—the fields were small and the marshes … well … marshy—but both were better than the forested areas that dominate the land around the airport. I was glad to have a plan before something happened. To this day, I still look at Google satellite images of airports I'm unfamiliar with so that when I fly into and out of them, I have some sense of what my emergency options might be.

The IMPOSSIBLE TURN

If you don't pick an area in front of you to attempt a landing, then you are probably considering the IMPOSSIBLE TURN. (Imagine a deep voice saying this with an echo reverberating for several seconds after the last syllable is spoken; that will give you a sense of the gravity and controversy about the topic.) The impossible turn generally refers to a turn made back to the field after an engine issue when below 1,000 feet AGL in a single-engine, light aircraft. It's called impossible because, in most cases and for most pilots, an attempt to make the turn ends badly. Most run out of altitude or put their plane into a stall.

What's interesting is that at altitudes well below 1,000 feet AGL (say between 500 feet AGL and 1,000 feet AGL) some pilots have demonstrated that the impossible turn is not impossible in some aircraft. But the key to making the turn and living to fly another day is this: you must fly very aggressively and fearlessly to successfully complete the turn. In fact, you must do something downright unnatural when you are worried about not having enough altitude: you have to get the nose very low, increasing your sink rate beyond that you would get with a best-glide speed and turn very aggressively, imparting G-forces to the wing that increase the stall speed (this, in fact, is why the nose needs to be low; you need to carry a lot of speed to accomplish the high-G turn without stalling). While in the turn, you can expect to hear your stall horn chirping at you like the tail of a rattler before the fateful bite comes. But only in flying this aggressively do you ensure you get your nose swung around quickly enough and with enough altitude to execute a landing.

For years, the FAA was dogmatic about not trying to return to the field at low altitude after a loss of engine power. A 2017 FAA publication (FAA-P-8740-44 if you're wondering) titled "The Impossible Turn" goes to great—and compelling—lengths to explain why the turn is *always* a bad idea. But just a year later, in Advisory Circular 61-83J, the FAA softened its approach, encouraging CFIs to "demonstrate and teach trainees when and how to make a safe 180° turn back to the field following an engine failure." The circular further notes that "Instructors should also train pilots of single-engine airplanes not to make an emergency 180° turnback to the field after a failure unless altitude, best glide requirements, and pilot skill allow for a safe return." You'll note that while this is a softening of earlier advice, it's not a suggestion that the turn is always a good idea. It's more of a "buyer beware" warning. It seems the FAA (understandably) wants to be sure the first time you attempt the maneuver is *not* when your engine has a power loss. If exploring the merits of the impossible turn are appealing to you, go up with an instructor (if they haven't taught you already) and attempt the turn, noting the altitude lost as you try to reverse course. Know the capabilities of you and your airplane. If you have the time and motivation to practice it periodically and the conditions on the day of your power loss are right, then maybe you'll have an option. I practiced the impossible turn a few times with an instructor at a safe altitude. I've decided it's not an option I would attempt. Maybe with more training and practice, I'll change my mind. But for now, I'm picking my best options out the front windshield.

Back in the cockpit, alone, I continued with my pre-takeoff briefing.

"When I have engine issues after takeoff and above 1,200 feet, I will pitch for best glide, turn into the wind, declare an emergency, and make my way back to the field, landing on any available taxiway or runway." Assuming there is some crosswind, and there almost always is, it may make sense to turn into the wind to begin your journey back to the field because it ensures that when you get your nose pointed back at the airport, the wind will be behind you pushing you toward runways and safety. Admittedly, you'd rather land into the wind, and you still may be able to if your altitude allows turns to position yourself in such a way. But even if you must land with a tailwind, that's almost always a better option than landing into the wind but off-airport.

That said, this is just a general rule. Depending on the layout of your home airport—especially if there are multiple runways—

and the strength of winds, it may make sense to turn in the direction where options will be off your nose the fastest. Because more taxiway and runway cement are to the south of Runway 29 at Hanscom, if winds are calm or light, I might decide to turn to the left (south) instead of into the prevailing wind which is usually coming from the north. There's often no one answer for every situation. *The point is to decide before you are forced to decide.* The less you must worry about in an emergency, the better.

Cleared for takeoff

With my emergency procedures soliloquy complete, I switched to the tower frequency and repositioned my airplane at the hold short line of Runway 29.

"Hanscom tower, November 238 November Delta, student pilot solo, short of 29, ready for takeoff."

"November 238 November Delta, Hanscom tower. Say intentions upon takeoff."

Damn, I thought, realizing I hadn't told him that I would be staying in the traffic pattern.

"Sorry," I said, sheepishly, "closed traffic."

"Roger November 238 November Delta. Hold in position. Landing traffic."

While the hold was a bit unnerving (I just wanted to *go*!), it was expected. It was rare to be given immediate clearance at such a busy airport. I could see a business jet on short final.

As I waited for the landing traffic, I did final checks. I had come to realize that there were many things I could forget or mess up while running through my checklists. I'd done it with Jeff. I'd even see Jeff do it. But I also realized that when it came to takeoff in my Warrior III, there were only a few things I had to be sure of:

1. The fuel selector was on a tank with fuel.
2. The fuel pump was on.
3. The mixture was rich.
4. The carb heat was off.
5. The flight controls were free and correct.

If just those five things were true, I could be reasonably sure I could execute a takeoff without incident. The first three steps ensure the engine has a fuel source (fuel selector), the fuel can get to the engine (fuel pump), and that it was being delivered to the cylinders in sufficient quantity (mixture). Step four ensures that the engine can deliver maximum power since carb heat reduces power.[40] And finally, the controls check ensures not just that the controls move when I move the yoke and rudder pedals, but that they move in the right direction. The second part of this check is critical. There have been accidents throughout history caused by a mechanic incorrectly reconnecting cables that run from the yoke or rudder pedals to the control surfaces. It's not enough to know that you can move the controls. You must ensure that the flight controls are behaving as expected when you do move them.

In a Piper Maintenance Manual, under the heading of a paragraph labeled "Aileron Controls Rigging and Adjustment," the following is written after the final step in a sequence related to how to check work done on control systems:

---

[40] Carb heat reduces engine performance because it introduces hotter—and therefore less dense—air into the engine. This has the effect of changing the fuel/air mixture in a way that robs the engine of power.

CAUTION: VERIFY FREE AND CORRECT MOVEMENT OF AILERONS. WHILE IT WOULD SEEM SELF-EVIDENT, FIELD EXPERIENCE HAS SHOWN THAT THIS CHECK IS FREQUENTLY MISINTERPRETED OR NOT PERFORMED AT ALL. ACCORDINGLY, UPON COMPLETION OF THE AILERON RIGGING AND ADJUSTMENT, VERIFY THAT THE RIGHT AILERON MOVES UP AND THE LEFT AILERON MOVES DOWN WHEN THE CONTROL WHEEL IS TURNED RIGHT; AND THAT THE LEFT AILERON MOVES UP AND THE RIGHT AILERON MOVES DOWN WHEN THE CONTROL WHEEL IS TURNED LEFT.

The bold and capitalized lettering of this warning and the plain-English language used to describe what *must* be done is a hint that this, like many procedures, regulations, and checklist items, is guidance that might save your life. Every time you get in a plane, you should assume it just came out of maintenance and the mechanic didn't follow these directions. Even if you own your plane and know with certainty it wasn't worked on, you should do the flight control check so that it becomes a matter of habit. Then, you'll be ready for that time when you will be flying a just-worked-on airplane. If you are in a larger flight school, you'll usually not have any idea if a plane has just come out of maintenance or not. So being regular about checking your controls is a must. One great trick I was taught regarding checking controls was this: when doing so in an aircraft with a yoke, point your thumb (or thumbs) up when your hands are on the controls. When you turn the yoke, your thumb(s) should point at the up aileron.

A small business jet landed in front of me and exited the runway.

"November 238 November Delta, cleared for takeoff, Runway 29, make left traffic."

"Cleared for takeoff, Runway 29, make left traffic, November 238 November Delta," I parroted back.

I turned on my landing light (which should be called a "landing *and* takeoff light" since it's used during both procedures to make your plane more visible to others), advanced the throttle forward, and steered the plane to the very end of the runway, using all available pavement. Runway 29 at Hanscom is just over 7,000 feet long. Given conditions that day, the airplane I was flying would probably get airborne in less than 1,000 feet. Still, I held fast to the old aviator adage that there are three things that are useless to a pilot: runway behind you, altitude above you, and one second ago.[41]

The left turning tendencies

As I lined up the Piper with the centerline of the runway, I checked my magnetic compass. Yep, 290 degrees. That matched the runway heading. Doing this brief and easy check ensures one last time that you have taken the runway the tower gave you clearance for. I got my heels on the floor to ensure I didn't inadvertently use braking to steer down the runway and then pushed the throttle all the way forward. As the engine came to full power, the airplane wanted to yaw left.

---

[41] Alternatively, I've also heard this as "runway behind you, altitude above you, and air in your tanks." Either way, the point is made.

The *PHAK* details well why the right rudder is necessary. There are four factors that cause the airplane to want to move to the left, particularly during takeoff. The first is the result of our old buddy Newton. His third law of physics, which Langewiesche leans on in *Stick and Rudder* as an explanation for why a wing produces lift, is at play. The propeller turns clockwise as you view it from the cockpit, so the airplane wants to turn in the opposite direction around the shaft driving the propeller—i.e., counterclockwise or to the left.

The second factor is the corkscrew effect. This is the result of how air moves behind a propeller in motion. As the name of this phenomenon implies, the propeller leaves behind it a clockwise-twisting, corkscrewed column of air lain on its side. It wraps itself around the plane like a boa constrictor wrapping itself around a dachshund. The direction of this corkscrewing means that a moving current of air strikes the tail of the airplane on its left side, pushing the tail to the right and the nose to the left (as you view the airplane from the cockpit).

Then there's the gyroscopic action. This one is most relevant for tail-dragging aircraft during takeoff. When your propeller is spinning, it is a giant gyroscope. I won't go into all the properties of a spinning gyroscope here, but the key driver of this left turning tendency is the fact that a force applied to a rotating disc perpendicular to the plane of rotation will cause a resultant force 90° from the point of the force and in the direction of the rotation. That's a mouthful, but it's not too complicated.

To picture this, imagine sitting in a tail-dragging aircraft during your takeoff roll. Your propeller is pitched back at an angle while the tailwheel is touching the ground during your initial acceleration. As you gain speed, you'll pitch the nose forward to raise the tailwheel off the ground. The propeller that was once pitched back is now (roughly) perpendicular to the ground. Now, briefly imagine the propeller as a rapidly spinning disk. Moving that disk from a "leaning back" position to a perpendicular one by lowering the nose of the airplane is no different than if you had simply pushed on the top edge of the spinning disk to move it from the angled position to the perpendicular one.

This is where the principles of a gyroscope come in. Pushing on the top edge of the spinning disk means that that force will be felt 90° from where the force was applied and in the direction of the rotation. Our force was applied at the 12 o'clock position, but it will be *felt* at the 3 o'clock position. And a push at the 3 o'clock position means that the nose will pull to the left. Voila, there's your gyroscopic effect.

Finally, there is asymmetric loading, also known as P-factor. Don't be daunted if you find the details of this one difficult to grasp. Even the *PHAK* says explaining this force is "complex because it would be necessary to work wind vector problems on each [propeller] blade while considering both the AOA of the aircraft and the AOA of each blade." Even with a math degree, I'm taking the FAA's word on this. That said, let me try to give a first order explanation by simplifying things as much as possible.

To start, it's important to note that the blades of a propeller are just tiny wings turned vertically. By spinning these tiny wings very fast, you get air flowing over an airfoil, and lift is produced. Because the direction of lift on these vertically aligned wings is forward, the "lift" pulls the plane forward. But that pull is not always uniform. When the airplane's nose is pitched up during takeoff and climb, the propeller blade descending between the twelve o'clock and six o'clock position has a higher angle of attack than the blade ascending on the opposite side (imagine the blade on the descending path taking a bigger "bite" out of the air). This higher angle of attack generates more lift and therefore causes more "pull" on the right side of propeller, yawing the nose to the left. That is your P-factor.

As I rolled down the runway the day of my solo (and still do today), I was conscious of all these left-turning tendencies. I reminded myself out loud, "Right rudder," to keep the plane aligned on the centerline and under control when it got airborne.

"Engine instruments in the green. Airspeed alive. Forty … forty-five … fifty … fifty-five. Rotate."

I think a lot of pilots underestimate the importance of the "airspeed alive" call. It's another habit you *must* develop if you are to be a safe pilot. Without an indication of airspeed in the cockpit, it can be exceptionally difficult to know how close you are to a stall or how to manage energy in the airplane through its various phases of flight. While there can be more than one reason for your airspeed indication to not work properly, often the issue has to do with the pitot tube. On a Cessna 172 it's the L-shaped tube under the left wing. On a Piper Warrior, it's a blade or fin-like structure under the left wing. Regardless of the shape, one common feature of all pitot tubes is a hole pointing forward, allowing air to flow into it, and another hole on the back of it to measure static pressure. If either hole is obstructed, you won't get an accurate (or any) airspeed reading in the cockpit.

This is why a thorough preflight is critical. A 2019 article on *Aviation Safety Magazine's* website details a study of accidents related to pitot tube issues. The report classified issues into four categories: contamination on the ground, in-flight contamination, malfunction of the pitot system, and failure to remove the pitot tube cover(s). Given how small the holes on the front of pitot tubes can be, it's understandable that in some cases ground contamination won't be visible. But an installed pitot cover (used, ironically, to keep ground contamination from fouling the instrument) should be as plain as the nose on your face. Unfortunately, the article notes, in 11 cases over a 19-year period, failure to remove a pitot cover was the cause of the crash. One of those crashes was fatal.

If you do notice an issue with your airspeed indicator after you're airborne, all is not lost. Pay attention to your visual cues and engine RPM. Remember the power settings and descent rates for your "perfect pattern" (we'll discuss what makes a perfect pattern in the next chapter). If you lock in on these measures when airspeed isn't available during your circuit around the field, you'll be okay.

As the airplane accelerated, I gently pulled back on the yoke and lifted N238ND into the air, surprised at how much more responsive the plane was without another full-sized adult in the cockpit. My solo had begun.

My first landing was perfect. My second so-so. I told the tower my third landing would be "full stop"; rain was threatening. I was happy with my landings and when I landed and was told by the tower to take the next taxiway to my left, I was proud beyond description when the controller said, "Nice job out there today." I'm sure he said that to all those who identified as "student pilot solo," but I chose to believe otherwise. Once off the runway, I switched back to the ground controller who had warned me of weather when I earlier called for taxi instructions, and she welcomed me back with similar enthusiasm. It was a great day.

I taxied back to the parking area at the west ramp. Jeff met me as I climbed from the aircraft. I retrieved the framed mementos from my father's shuttle mission from the cargo compartment of the Warrior and stood in front of the plane holding it while Jeff snapped a few photos of me.

Shirttails and ... spankings?

For most student pilots, the completion of their first solo culminates with an unusual ritual: cutting off the pilot's shirttail. The excised portion of the shirt is then inscribed by a permanent marker with the date of the solo and the airport where it was completed (and any other meaningful information the pilot and instructor want to record for posterity).

The explanation of the origin of this tradition is consistent across sources. In the early days of flight training, when instructors were usually sitting behind students in tandem-seat biplanes with open cockpits and no communications headsets, instructors would tug on the shirt tails of students when trying to get their attention. Upon gaining that attention, instructors would then point in a direction for the student to fly or shout some instruction at them. After a first solo, when the student had demonstrated the ability to act as pilot in command of their airplane, there's no need for a shirttail anymore because there's no need for an instructor to get the student's attention.[42] Might as well cut it off. I was a bit disappointed that the strange practice of cutting off shirttails wasn't common at my school, but I must admit that I was glad to keep one of my favorite workout tees intact.

As bizarre as the cutting of shirt tails custom is, there are stranger ones. In one article I found, it was noted that in

---

[42] I'm assuming that in the "old days" once a student soloed, they were considered a pilot and additional instruction wasn't necessary.

Germany a common commemoration of a first solo involves student pilots bending over the back of the airplane and letting instructors spank them in a sort of "welcome to the flying club" hazing ceremony. I tried to find pictures online of such a moment. While on a lunch break at work I typed into the Google search bar "solo spanking German pilot." Bad move. While I didn't click on any links, I didn't get the impression that any of them dealt with aviation matters. If I'm unemployed by the time this book is published, you'll know why.

Chapter 13
Landings Part I: Pilots Manage Energy

Flying is an activity that is all about dancing with the rules of physics. From the moment you start the engine of your plane, the laws of nature and science both facilitate what you are trying to do and bound what is possible. The air will provide you with lift, but only if you keep the wing's angle of attack within limits. The engine will produce enough power and the wing enough lift to get you airborne on a short runway, but not if density altitude and aircraft weight get in the way. Gravity and friction will help you to land, but too much of either can end badly. As Langewiesche noted, flying is an activity of contradictions.

Flying successfully in that world of contradictions is all about energy management. In fact, at its core, that's what a pilot's job is: to manage energy. Do that well and consistently, and you will be a safe and impressive aviator. Do it poorly and you may be calling your insurance company to see if the damage is covered.

Conservation of energy

To better understand what it is you're trying to manage while flying, it's helpful to understand a few basic principles of physics.

The first—and perhaps most important for the purposes of our discussion—is this: energy cannot be created or destroyed. It can only be converted from one form into another.

The second is this: there are multiple forms of energy. You may remember from your high school physics class that if you carry a ball to the top of a hill, you impart to it something called gravitational potential energy. You essentially are storing energy in the ball by placing it at a height. When you let go of the ball and it rolls down the hill, that gravitational potential energy is converted into *kinetic* energy—the energy related to motion. In an imaginary world with no friction due to the surface of the hill or the air, that ball would roll down the hill and make it all the way back up to the top of an equally high hill on the opposite side of the "valley" between the two prominences. This is because of the first thing we mentioned. Since energy cannot be destroyed, all the energy that existed at the beginning of the ball's roll must exist at the end in some form. In fact, throughout the ball's journey in this example, its energy state is a zero-sum game; when one energy goes up, the other must go down by an equal amount.

So it is with an airplane. When your airplane has altitude, it too has gravitational potential energy. You can convert this stored energy into kinetic energy (airspeed) by losing altitude. From an aviation physics perspective, this is the equivalent of letting a ball roll down a hill. Likewise, if you have kinetic energy, you can convert this back into gravitational potential energy by gaining altitude. This is our ball rolling back up the hill. The higher the altitude, the more gravitational potential energy you have.

I noted that in our thought experiments above, we assumed there was no friction involved. In reality, of course, this isn't the case, for the ball or for your airplane. The ball is affected by the friction of the ground and the friction of the air around it. When airborne, the airplane generates heat energy through air friction (the pull of the atmosphere trying to slow it down, a.k.a. drag). It also generates sound energy through vibrations in the engine and frame of the airplane and atmospheric kinetic energy due to air being moved by the plane flying through it. All these forces "steal" some of the energy from the plane's kinetic and potential energy. That's why, if you were flying at 3,000 feet, shut down your engine (don't do this!), dipped the nose 10°, and lost 1,000 feet, gaining speed as you went, then pulled up 10° and tried to gain those 1,000 feet back, you wouldn't quite make it back to where you started. Some of your potential and kinetic energy has been converted into heat energy, sound energy, or energy in the form of moving air. Your roller coaster has lost some steam.

Speaking of roller coasters, they are a great (and common) example of the principles discussed here. A chain pulls the roller coaster to the top of a large hill and then lets it go. Potential energy is turned into kinetic energy as the coaster screams down the first hill. In most cases, the coaster will then journey up and down a series of additional hills. But you may have noticed that the hills get smaller and smaller. This is because the loss of energy through air and rail friction, noise, air movement, and—at the end of the ride—brake friction ultimately bleeds the coaster of all its potential and kinetic energy. If these losses didn't exist, the coaster could (and would!) go on forever.

Unlike a roller coaster, we can overcome the theft of energy from our plane and keep it flying due to another form of potential energy at our disposal: chemical potential energy. The fuel in your airplane's fuel tanks is energy just waiting to be released. When we burn that fuel, we convert it into kinetic energy—in the pistons, camshaft, crankshaft, and propeller. The addition of power allows us to climb back up that imaginary hill to our original altitude in the example cited earlier. Or, if we decide to stay straight and level, more power can be converted into airspeed. It's also common for the middle ground to exist: you can gain both airspeed *and* altitude depending on your engine performance and climb rate.

I know that all of this can seem very daunting. And understanding the interplay between different forms of energy in an intellectual sense is not necessarily critical to being a good pilot. But understanding it *practically* is. And it's most important when you are landing.

Energy management when landing

It might rightly be said that anybody can "fly" a plane. But not everybody can land one. Indeed, landing is where you will spend a disproportionate amount of your training time. And, unfortunately, landing is universally regarded as the hardest thing to master. That's why, like many student pilots, I could have nailed every steep turn, stall recovery, turn around a point, and simulated instrument scenario on a training flight, but if I bounced, floated, or dropped when landing, I drove away from the airport feeling like a field goal kicker who missed a game-winning kick as time ticked off the clock. Unfortunately, I often got in my car after a lesson feeling exactly that way—and still do on some days even now! You will too.

It's not a surprise landing is difficult. In fact, when you think of all that's going on, it's amazing that any landing is a "greaser." You have to take a relatively fragile machine, decelerate it from speeds that most of us have never attained in a car, drop from a traffic pattern altitude equal to the height of a 100-story building, adjust the airplane's path based on continuously changing winds, and coax it onto the ground no more than inches left or right of a centerline, dissipating forward and vertical energy so that the plane—and you—don't even feel the touchdown. That is *hard.* I don't care who you are. Guiding an airplane to a feather-soft return on the runway depends on your ability to bleed energy from the airplane so that when the wheels touch the ground and the gravitational potential energy has gone to zero, the kinetic energy is manageable.

My first real try at executing a landing was at the end of my second flight. It didn't go well. Jeff certainly didn't say that; he was always encouraging. But I knew it wasn't up to standards. Many of my subsequent ones weren't either. I tended to come in fast and float way too long. I wasn't managing energy well. Not a huge issue on a 7,000-foot runway (although still risky), but it's problematic when trying to land at the thousands of much shorter and narrower runways around the country—and it's a lack of precision that would not be tolerated during an FAA checkride.

"Do you think you are too low or too high?" Jeff asked me one day when we were about one mile from the runway on final approach.

"Too high," I answered. It was easy to tell. The PAPI (Precision Approach Path Indicator) showed four white lights and no red ones, meaning I was above the proper glide slope[43] to

---

[43] Glide slope can have a specific definition related to the use of technology to aid a pilot in landing when they are in or above clouds or approaching through fog. In this book, I use it more generally to mean the proper location of the airplane in both altitude and position over the ground when lining up to land.

the runway.[44] PAPI's are a great tool when landing (as are VASIs—Visual Approach Slope Indicators). The lights to one side of a runway help ensure you are on the glide slope to land somewhere around 1,000 feet from the end of the runway. I found them particularly helpful when learning to land. Even though every airport won't have them, if you get to train at a field with them, they help you get an understanding of what the view out the window should be when you are properly aligned during final approach.

Jeff certainly knew I knew I was too high. His question was an indirect way of asking me what I was going to do about it. This was an important lesson. "Staying ahead of the airplane" is something I guarantee you'll hear repeatedly during your training. It means not letting things that can be addressed now go unaddressed, making them harder to deal with later. To be clear, this *doesn't* mean frantically and abruptly acting in a way that makes things worse. For example, being above the preferred glide slope while landing is something that must be fixed with finesse. Cutting the power and pitching the nose of the airplane down is not the way to correct the problem (more on this momentarily) and it may have you blow right through where you

---

[44] PAPI lights are typically in a row—four lights across. If you see two red lights and two white lights, you're on the glide slope. If you see more red than white, you're too low. If you see more white than red, you're too high. A VASI usually has two lights in one row and two lights in another positioned along the side of the runway and in line with each other. This gives the impression on approach of two lights "over" two others. The principle is the same as the PAPI—you are looking for red lights over white; this means you're on glide slope. If you see "red over red" then "you're dead," meaning too low. If you see "white over white" then "you're high as a kite," meaning too high.

should be, putting you on the low side of the glide slope. Over-correcting back the other direction could then have you "porpoising" through the glide slope in an uncontrolled manner. Act with purpose but act with discipline.

Speaking of discipline, looking back now, I realize that early in my training I was not disciplined about all the things that would increase the likelihood of executing a great landing *substantially*. To this point, there's an old adage I can't emphasize enough:

The perfect landing begins with a perfect pattern.

When I first heard this, I thought too literally about it. When we think of the word "pattern," we think in terms of shapes. Of course, a nice rectangular pattern made up of departure, crosswind, downwind, base, and final legs with opposite sides of the rectangle equal in length is one of our objectives. But as important as the shape of the pattern you trace in the sky is, it's what's happening in all the other areas I touched on earlier that matters most.

First, let's talk airspeed. This is an area of common difficulty. Most trainees fly the pattern too fast. It's not hard to come up with a hypothesis as to why this is the case. So much of what we student pilots have drilled into us on the ground and in the air relates to slow flight and the dangers of a stall, particularly when low. Fear of a stall leads pilots to add "buffer" to airspeeds in the pattern. If your final approach speed should be 65 knots, then surely 75 knots is safer, right? A 2022 article by David St. George on the Society of Aviators and Flight Educators (SAFE) website describes this issue well. The article is called "The Dangers of Timid Piloting," and in it St. George discusses several problems he sees related to pilots not being aggressive enough in how they fly based on a misunderstanding of aerodynamics. A timid pilot, he says, "adds more 'airspeed buffer' and flies way too fast in the pattern. This leads to being unable to slow down and stabilize the final approach for a normal landing. All the accumulated energy gained through aerodynamic ignorance creates a much more dangerous landing."

I was guilty of exactly this. For the better part of the first half of my training, I flew my approaches much too fast. This led to significant float, longer landings, bounces, and the like. I wasn't managing the atmosphere's theft of energy from the airplane through drag well enough. This makes the possibility of an unfortunate event like a prop strike (the prop hitting the ground) or loss of control while on or near the runway much more likely—no matter how long that runway is.

The solution, obviously, is to fly the approach at the right speed. But how?

Power in the pattern

Defining power settings for different segments of the pattern does a *lot* to ensure your speed is appropriate for a successful landing. Not only that, but the right power settings will also help ensure you are at the correct altitude at the right time for a particular aircraft configuration (e.g., with flaps fully deployed). When all these variables are within an acceptable range during your landing attempt, you are said to be on a "stabilized approach," a term you are likely to hear during your training. The good news when considering power settings is that (for most trainers) there are only two such settings that matter while in the pattern. The first is what you set your power at during the first part of your downwind leg after you've attained pattern altitude.

In the airplanes I was flying on most days, that equated to about 2,100 RPM. The second is when you put in your first notch of flaps abeam the runway numbers. On most days that was around 1,800 RPM. (Note that I say "most days" because these settings change based on engine performance differences driven by, for example, higher or lower density altitudes.)

As I roll into my downwind from my crosswind at traffic pattern altitude, a power setting of around 2,100 RPM with no flaps and trimmed for straight and level flight will put me at about 90-95 knots.[45] When I pull the power to around 1,800 RPM and put in one notch of flaps when abeam the numbers and trim the nose down so I have a descent rate of 500 feet per minute (fpm), the plane will slow to about 80-85 knots. Here's the beautiful thing: when I turn base and put in my next notch of flaps and trim to maintain my 500 fpm descent rate, the plane will further slow to about 70-75 knots *without me touching the power.* Finally, three notches of flaps on final with that same descent rate will yield about 65-70 knots until you make your final adjustments in your roundout and flare (more on this later). Of course, this assumes everything is perfect, which it never is. Still, having these power and speed benchmarks in your head and aiming for them is critical if your landing is to be a good one.

The other thing that took me a while to internalize when it came to airspeed and concern about stalls while landing is this:

---

[45] "Trim" here refers to the use of a small control surface (called a "trim tab") on the elevator or stabilator to hold the aircraft at a particular angle of attack for a given power setting. You usually set the trim by way of a vertical wheel or (if electrically enabled) through a control on the yoke. Trimming an airplane correctly helps to relieve the pilot of control inputs to get a desired attitude. Aircraft can also have rudder and aileron trim inputs, but the elevator is by far the most commonly available and used.

once you get abeam the numbers and put in that first notch of flaps, you should be in a constant descent to the runway with your nose below the horizon. While the location of your nose doesn't guarantee you are staying away from a stall—remember, you can stall an airplane at *any* airspeed and *any* attitude—if you are flying your pattern with the correct flap and power settings at the correct time, you won't get the wing past that critical angle of attack if you are descending.

Even though your nose will be low, you will be slowing as you add flaps. The use of flaps during the latter part of our pattern introduces more lift, which helps the plane keep flying at slower speeds. One consequence of this increased lift is more drag. The drag helps us to lose gravitational potential energy (because we are losing altitude), without converting *all* of that lost potential energy into kinetic energy in the form of airspeed. That's because some of that energy is being transferred (remember—it can't be destroyed!) into heat and kinetic energy in the air through the incremental drag the flaps create.

Pitch and power, speed and altitude

So, you've flown the perfect pattern but realize that even when you do, you'll need to make adjustments when on final approach to keep yourself on the glide slope. Maintaining the proper angle of descent to the runway and at the right airspeed requires you to adopt an approach to controlling the airplane that I'm sure you've heard from your instructor: pitch for airspeed, power for altitude.

But wait, isn't this backward? Wouldn't I point the nose down if I want to go down? In other words, pitch for altitude? Wouldn't I push the throttle forward to go faster? It's indeed true that if you pitch your nose down, in most circumstances, you will lose altitude. The problem is that you will also, in most circumstances, gain airspeed because of that pesky conservation of energy issue. Likewise, if you add power, you might go faster, but only if you keep the nose level or down. When you are trying to land, doing what seems obvious can have secondary effects—higher airspeed, higher altitude—that are the opposite of what you want.

So instead of doing what feels intuitive, we flip the world on its head and use pitch to manage airspeed and power to manage altitude.

To explain further, I want to go back to the types of energy at play while flying. Imagine four buckets that get filled and drained with each kind of energy during your flight. There's a kinetic energy bucket, a gravitational potential energy bucket, a chemical potential energy bucket, and a bucket which, to simplify, I'll just call the "drag" energy bucket. The contents of the drag energy bucket represent the energy that has been "stolen" from the airplane and transferred into the surrounding environment by its movement through that air and across the ground (during taxi, takeoff, and landing).

With those buckets in mind, recall that energy cannot be created or destroyed. That means that in an imaginary bubble around your plane, the total amount of energy represented by the cumulative contents of each bucket *never changes.* If the amount of energy in one bucket goes up by *x* units, then one or more of the other buckets *must* see a decrease in their contents that totals *x*.

When the airplane is parked on the tarmac with the engine off, all the energy is in your chemical potential energy bucket. The plane is not moving, so it has no kinetic energy. It has no altitude above the surrounding ground, so it has no gravitational potential energy (presuming it's not parked at the top of a big hill!). And because it's not moving, there is nothing in the drag bucket. Now, fast-forward to takeoff. You push the throttle forward. As you do, you burn fuel and therefore decrease the amount of energy in your chemical potential energy bucket—in fact, this is the easiest energy bucket to visualize because it's literally the fuel in your fuel tanks. But that chemical potential energy doesn't just disappear, it flows into the other buckets. As you accelerate down the runway, but before you get airborne, your kinetic energy bucket begins to fill as your speed increases. More speed through the air means more drag, so your drag bucket will also begin to take on some energy. Your gravitational potential energy bucket is still empty since you haven't gained any altitude yet. But when you lift off, that bucket too will begin to fill. The gravitational potential energy bucket will continue to fill until you level off, at which point the energy from your chemical potential energy bucket will continue to go down as you burn fuel to overcome drag and gravity, your kinetic energy bucket will fill a bit more as you accelerate to cruise speed, and your drag bucket will continue to fill as it steals energy from the plane. And so it goes throughout your flight; buckets fill, then drain, then fill again, but the total amount of energy across all buckets stays the same.

When you ultimately land, you are trying to "bleed off" as much of the kinetic and gravitational potential energy as you safely can by filling your drag bucket. This is where our "pitch for airspeed, power for altitude" philosophy comes in.

Here's why. First, understand that your chemical potential energy in the form of fuel in our fuel tanks is energy that you *can* bring back to the ground with you. You don't have to "bleed it off" like your gravitational potential energy and your kinetic energy.[46] These latter two energies *must* be dissipated if you are ever to accomplish a safe landing and ultimately stop on the ramp when you park the airplane. (In fact, the definition of a parked plane could be "the point at which the kinetic energy and gravitational potential energy in the airplane are equal to zero.") When you leave the power alone and pitch the nose down to gain airspeed or raise the nose to lose airspeed, you haven't added to the combined value of the potential and kinetic energy buckets. You simply traded something in one of those buckets for something in the other while allowing friction to slowly drain energy from both buckets. The minute you burn fuel, *you add to the buckets* because burning fuel will either increase your altitude or increase your speed (or both).

So, when do you add power? Only add power when you *must.* And when is that? It's when you want to convert some of your chemical potential energy into gravitational potential energy by increasing the altitude of the airplane to ensure that you make it to the runway.[47] This works because an airplane trimmed for level flight will, if power is added, climb; it won't gain speed. In fact, here's what's cool: it will climb at roughly the same airspeed it was at in the lower power setting. And it doesn't matter how much power you add. The plane will naturally pitch up to a

---

[46] It is true that carrying around fuel requires you to have more energy in the airplane. This is because a heavier airplane needs more energy to fly. So, in this sense, part of the energy you'll have to bleed off from the plane is there because of the fuel weight you are carrying around.

[47] Some might say you also add power when trying to increase speed to stay clear of a stall. Sort of. But remember, we'd like to use pitch to control airspeed. If you are too low to dip the nose to gain that airspeed, adding power helps you gain altitude so that you *can* then dip the nose to establish a safe speed.

degree that will maintain the airspeed you had trimmed for before you applied more power. So, by adding power when you are low, you increase your altitude and your potential energy, but without immediately converting it into kinetic energy; your speed stays the same. This is very useful when you are trying to keep your airspeed in a manageable range during the critical moments before landing.

When we take power out, the physics are the same. The airplane will want to descend at roughly the same airspeed it was trimmed to when power was in. In both the ascending and descending cases, the speed will only stay stable though if you don't pitch the airplane with the yoke at the same time. Pushing the yoke in or pulling it back will change your speed—this then is how we "fine-tune" airspeed while landing.

I mentioned that when you park on the ramp your kinetic and potential energy buckets will be empty and your chemical potential energy will have been reduced. Where did all that energy go? It went into the air around you as friction turned to heat. While this conversion was happening throughout all phases of flight, it became more pronounced as you dropped flaps, exposing more of the oncoming air molecules to the surfaces of the plane. It became more pronounced when your wheels hit the runway and began rubbing against its surface. It became more pronounced when you engaged the brakes and the calipers clamped down on the wheel rotors. Nature is continuously trying to steal energy from the plane. For most of your flight, you fight this by running the engine fast and keeping the airplane "clean" (no flaps, gear up, etc.). But when you land, you let nature have what it wants, in just the right doses, draining your kinetic energy and gravitational potential energy buckets until they are empty.

I hope this explanation helps you understand the reason behind our "pitch for airspeed and power for altitude" approach to landing. No doubt, it takes some mental energy to understand. If you struggle with the details, a simplified way to consider this method of energy management is to think of your elevator as your gas pedal. Push forward to go faster, pull back to slow down. Your throttle is your altitude lever: push it in to go up and pull it out to go down. As with other aspects of flying, I've found that reminding myself out loud, "Pitch for airspeed, power for altitude," while on final is helpful. For me, this was something that didn't make sense on one landing but then did on the next. It "clicked" in an instant.

Chapter 14
Landings, Part II: It's All About the Aim

All of this is great, you may think. *But what you've taught me only gets me to short final. How do I get the airplane on the ground?*

Indeed, this is the big question. The last 15 seconds of a landing is where the rubber meets the road (or runway, as it were). You're lined up on final, your airspeed is looking good. You're descending on the glide slope. Your "pitch for airspeed, power for altitude" mantra has got you locked in. But how in the world do you execute the final steps so that your wheels touch down where you want them to with a sweet, gentle chirp?

Advice abounds when it comes to how to answer this question. But I've found a few nuggets that really seem to work.

First, have an initial aim point. By that I mean have a place on the runway along the centerline you are focused on in the final phases of your descent. I usually use the runway numbers. Keep in mind this isn't the location you will land. At this time in your approach, your nose will be pointed down relative to the horizon. The aim point is what will guide you to the place where you will begin to lift the nose to arrest your descent.

As you approach the runway, the aim point should be fixed in your windscreen. In other words, if you were to mark the inside of the front window where the aim point was fixed in your view (some instructors may indeed actually do this), that mark would stay on the aim point. The reason this is helpful is because it indicates that the plane is on a defined descent path and aligned with the centerline of the runway. Keep in mind though that achieving this alone isn't sufficient for a successful approach. After all, you could be at 2,000 feet just a quarter mile from the end of the runway, have the nose pointed nearly straight down, and have your aim point fixed in your view (you'd essentially be a dive bomber in this scenario). This won't work; your excessive nose-down attitude will result in very high airspeed. No, in addition to having the point fixed in your view, you *also* need to have your altitude right. This is why landing is difficult. If you could just do one thing—get a point fixed in your forward view—and that led to a greaser landing, we'd all be experts (and the cost of training would be cut in half). But unfortunately, that's not the case. You need have your aim point stable in your view *and* be at the correct altitude and, therefore, the correct speed. Doing these things means you will be on the glide slope.

As I noted earlier, there will be a point at which you will need to transition from your nose-down attitude to a nose-level attitude by applying back pressure to the yoke. If you locked in on your aimpoint and remained locked in on it until the moment the plane met the ground, the NTSB report would describe your crash as the result of CFIT (pronounced SEE-fit): Controlled Flight Into Terrain. We don't want that; *that* will not impress your instructor.

There are many videos and blog posts that address the topic of when to begin what most call the "roundout"—the phase of a landing profile when you are transitioning the airplane from nose down to nose level and then nose up for touchdown, slowing its forward and vertical speed along the way. Many of those sources of information will describe the final phases of landing as two distinct parts: the "roundout" and the "flare." Among those that hold this opinion, the roundout is a leveling of the airplane to bleed off speed and the flare is a distinct pitching up of the nose to ensure (in a tricycle landing gear airplane) that the rear wheels touch the pavement before the nose wheel while the plane is at or near stall speed. Landing with this orientation is important to ensure you don't damage the nose wheel and/or allow the prop to hit the ground, not to mention that it helps dissipate that last bit of remaining energy to ensure a safe, slow touchdown. However, the FAA's *Airplane Flying Handbook* uses only the term "roundout," there are not two distinct phases in their description of a proper landing. It's a single, continuous maneuver.

The FAA's definition notwithstanding, I *do* think of the roundout and flare as two distinct phases. Doing so helps keep me from raising the nose too soon. The roundout is an opportunity to "settle" a bit—to allow the plane to arrest its glide slope sink rate to a slower one. It helps me think of the landing in discreet pieces, even if somebody viewing the landing from the tarmac may see it as one continuous transition.

Much has been said and written about when to begin your roundout. The best advice I've found is to begin it when the edges of the runway start to fill the edges of your field of view (i.e., the sides of your front windscreen) rapidly—when the trapezoidal shape of the runway (it's that shape because of your perspective when landing) becomes flatter and fatter so that the sides of the runway trace lines that touch the sides of your instrument glare shield. Be ready for this to happen quickly and recognize the speed in the change of view as a clue that it's time to arrest your descent. The speed with which the view out the window changes can surprise you because the relative size of the runway from left to right doesn't expand at a constant rate as you get closer to the ground. Rather, when you are about 10 feet above the pavement, it will grow in your front view very rapidly—about 10 times faster than when you are higher in altitude earlier in your approach.

Note that this method only works well if you are on speed and on glide slope. If these two conditions are met, when the runway expands as described, you can pull your power to idle and gently pull back on the yoke to level the airplane. This will have the effect of forcing your eyes to the horizon, something you want during the final phases of landing as it helps you better judge where you are relative to the ground.[48]

As the airplane begins to "settle"—i.e., slowly lose altitude—pull back farther on the yoke. Doing this raises your angle of attack, increases lift, and increases drag, helping to slow the airplane's descent while also slowing its airspeed. At this point, you have very little gravitational potential energy, and what you do have will continue to decrease to zero as you get closer and closer to the ground. Your kinetic energy is also decreasing as friction takes its toll. You're setting yourself up for a smooth landing.

---

[48] This helps you gauge altitude because during other phases of flight when on the ground—namely taxi and takeoff roll—you are looking forward toward the horizon. Therefore, whether you are conscious of it or not, you have an expectation of what the view out the front window should be when you are very near the ground.

Ballooning and sinking

The description in the previous paragraphs make all of this seem easy. It's not. Even if the pattern, descent, roundout, and flare are perfect, you're still not home free. It's a rare landing that my pull on the yoke is smooth and continuous all the way until touchdown. Sometimes, the plane may want to "balloon" up as I pull on the yoke. Sometimes, I lose airspeed too rapidly and the airplane wants to sink abruptly. The former may be because I'm too fast. The latter because I'm too slow. But even if I fly the approach perfectly, gusts of wind—or the sudden loss of wind—can intrude, making the airspeed (and the plane) bounce all over the place. Coaxing the plane to the ground can become a challenge.

If this happens, it's important to compensate. Always keep flying the airplane; don't let things happen *to* you. Rather, *decide* what will happen. Act with confidence but also with finesse. If, for example, the plane balloons—meaning it arcs upward as you pull back on the yoke because you have too much airspeed and that airspeed is being converted into altitude (conservation of energy again!)—the worst thing you can do is aggressively lower the nose by pushing forward on the yoke (unless you are near a full stall high above the ground). Doing this takes your new potential energy in the form of higher altitude and converts it into kinetic energy. Your airspeed will increase. As you dive at the runway at a higher speed, your instincts will be to pull back on the yoke. But again, all you will do is convert that speed back into potential energy. You can end up porpoising down the runway.

The best approach if you do balloon up when you pull back on the yoke and you have enough runway and enough airspeed, is to make sure the power is at idle and keep the nose level to slightly up. Let the higher angle of attack and increased friction cause the loss of gravitational potential energy and kinetic energy. Reestablish yourself in the beginning of a roundout. When the speed decreases sufficiently and the plane begins to sink, begin pulling back on the yoke raising the nose. Of course, all of this assumes you are close enough to the ground and have enough runway left to make this doable. You don't want to "land" while you are still 30 feet above the runway. We don't call this a landing. We call it a crash. And you don't want to "settle" the airplane as the last few feet of runway pass underneath you, forcing you into the perimeter fence. We also call this a crash.

In the case of too little airspeed, you may feel the airplane fall out from underneath you as you raise the nose in your flare. This is a sickening feeling. You'll find yourself wincing as you wait for the stall warning horn to sound, looking like a bomb diffuser fearing an incorrectly cut wire. In this case, more power may be needed. Remember that our chemical potential energy is there for a reason—to increase our total energy state by converting the chemical energy into kinetic and/or gravitational potential energy. By adding power while keeping the nose level, you'll add kinetic energy and lift, helping to arrest a sudden drop.

In the scenarios above, I've talked about how to compensate for a less-than-perfect final few moments of the landing process. But you should only be compensating in such a way if you entered the landing in a stabilized fashion. In other words, you've given yourself the best opportunity to land safely. But keep in mind that if things aren't looking great, you *always* have at your disposal the go-around. It's likely the most underutilized of all maneuvers.

The go-around

Very early on in your training you will have drilled into you that a go-around is always available to you. "If something doesn't seem right on landing," you're told, "use your engine. That's what it's there for."

It *is* one of the things the engine is there for. Every pilot should use it if necessary. The fact of the matter is, though, few do. This is even true of professional pilots. In a 2017 study by the Flight Safety Foundation called "Go-Around Decision-Making and Execution Project," authored by Tzvetomir Blajev and William Curtis, it was found that "failure to conduct a go-around is the number one risk factor in approach and landing accidents (ALAs)" in commercial operations. The authors note another study that showed that between 1994 and 2010, unstable approaches made up 3.5 to 4.0 percent of all approaches and that in 95 to 97 percent of those approaches, "flight crews whose airplanes were in this state continue[d] the approach and landing." Further, again citing previous studies, the authors noted that "approximately 65 percent of all accidents are ALAs" and that "one analysis concluded that more than 80 percent of ALAs would have been preventable—if the crew had decided to go around."

What's true of commercial pilots is doubly true of general aviation pilots. We are, I think, even more reluctant as a group to initiate a go-around. Why? Almost certainly a big part of it is ego. Pilots tend to have healthier egos than the general population and, whether they know it or not, might internalize the old fighter pilot mantra, "Better dead than look bad." To the extent that some may view a go-around as an admission of failure because it implies the approach was not perfect, this attitude can be seriously dangerous. Secondly, at a busy airport, doing a go-around might feel like an inconvenience to other aircraft or air traffic controllers. You've upset their finely tuned plans and you've also added time, cost, and stress to your flight.

I think it's also possible that many pilots are simply afraid of go-arounds, viewing them as dangerous maneuvers. You are low to the ground. You are slow. You have flaps extended. All these factors—along with the forces the plane will experience when you apply full throttle and try to climb out—may lead pilots to make inaccurate judgments about relative risk. In other words, pilots may convince themselves that while their approach is unstable and there is risk in that, a go-around has risks as well and those might be higher. While there are some scenarios where this may be true, it's highly unlikely that in all the scenarios where a go-around was *not* selected the risk of the go-around was greater than continuing the approach.

One reason all these factors are at play is that go-arounds are just not practiced enough during training. I don't think I did more than three in seven months. It's my belief that many instructors, when thinking about how to use precious training time, bias themselves to using it for an actual landing. After all, when you do your FAA checkride, it's guaranteed that you'll have to demonstrate your landing ability. You won't necessarily be asked to demonstrate a go-around (although it's important to remember that during your checkride you can always initiate a go-around yourself if your setup for landing is not ideal).

While go-arounds can be the safe choice, it *is* also true that they can be hazardous if you don't execute them properly. Three things—or a combination of them—get pilots in trouble when trying to execute a go-around. And each has to do with a nice sing-songy phrase that you should have in the back of your head when executing a go-around: power up, pitch up, clean up. Do these things correctly and all will be fine. Do them incorrectly and things may not turn out very well.

First, let's discuss "power up." There can be a tendency to jam the throttle forward too quickly when deciding you need to do a go-around.[49] Doing this introduces the forces we discussed earlier that lead to the left-turning tendencies. But the onset of those forces may be abrupt to the point of surprising. Your engine will usually be at a low power setting—often at idle—at the point at which you decide to do a go-around, and so you probably won't have much right rudder applied. Since adding power and pitching up in quick succession will lead to the onset of all the left-turning forces, you need to be ready on the rudder. If you aren't, you risk losing control of the aircraft, especially if you are flying a plane with a powerful engine. Remember that in most cases you don't need to smash the throttle forward. You want to apply power expeditiously but doing so in one smooth motion is usually sufficient and safer. Whenever you are doing something that you don't usually do or dealing with an emergency, your mindset should be that of a U.S. Navy SEAL: *slow is smooth and smooth is fast.* Doing things in a panicked rush rarely makes the situation better. A go-around is no different.

The second potential issue with a go-around has to do with the next step in our ditty: pitch up. When you decide on a go-around you are most likely slow. Even if you are "fast" in the sense that you are too fast to make a safe approach and landing (and that's why you want to initiate a go-around), you are still

---

[49] This tendency to jam the throttle forward is one reason I don't like the "3 Cs" some people use when abbreviating a go-around procedure: cram, clean, climb. Cram revers to "cramming" the throttle and mixture forward. "Cramming" sounds a bit too panicked to me.

slow relative to cruise flight. It's common for an inexperienced pilot to want to yank back on the yoke as they add power. This is a bad idea. Doing so increases the angle of attack before speed has been gained. This puts you at risk of a stall. Not good. In a large majority of cases, you have the time to keep the nose level as you apply power and gain speed. Do this; it doesn't take long. In just a few seconds or so you'll be in a good spot to safely climb back to pattern altitude and give it another go.

Finally, the "clean up" third of our three-part go-around phrase comes into play. By "cleanup" we mean getting the gear and flaps retracted to reduce drag on the plane. For most of us early in our training, gear isn't an issue. It's permanently down. But flaps are another story. In a 2007 POH for a Cessna 172, a go-around is mentioned with distinctly highbrow language. It's called a "balked landing." This sounds like language Lord Haldane might have used. In the procedure under the "balked" title, it notes that the pilot should immediately raise flaps to 20°. This is the second notch—the first notch being 10° and the third being 30° (in later 172 models). While flaps increase lift, they also increase drag, so raising them is necessary if you are to effectively gain airspeed and climb back to pattern altitude. This is why raising the flaps that first 10° is so critical. But many pilots will make the mistake of raising the flaps *completely* in a rush to do a go-around. This can cause the plane to abruptly lose lift before it's got enough airspeed to climb and result in the airplane falling out from underneath you. Depending on your speed and altitude, this can be anything from an adrenaline-inducing annoyance to a fatal mistake. Like other parts of our "power up, pitch up, clean up" ditty, it needs to be done with another ditty in mind (my own): "quick and confident, but controlled."

Fly all the way until touchdown ... and parking

If your approach is looking good and a go-around is not necessary, keep pulling back on that yoke as you get closer and closer to the ground. Usually, you should hear your stall horn going off, getting louder and louder as if it's getting more and more excited about the wheels of the airplane touching the pavement. The noise of the stall horn tells you that you are about as slow as you can possibly go. This is the only phase of flight where being that slow is a good thing. Being slow and *very* low—like six inches low—is what you want during a landing.

Once those main gear wheels hit the ground, don't get complacent. You still have some flying to do. Winds are still acting on your aircraft. You still have energy in the plane as you roll out to exit the runway. Others may be getting ready to land behind you. Stay focused. It would be a shame to waste your perfect execution of a landing by running your plane off the runway because of a crosswind, or taking a wrong turn upon exiting the runway, or trying to execute your after-landing checklist while you are still moving. You don't power down your brain when you are driving a car, I hope. Don't power down your brain while "driving" an airplane on the ground.

Chapter 15
Cross-Country Planning

"I'm heading to the airlines," Jeff said to me one day when I arrived for a lesson.

"Awesome! Congrats!" I said. And I meant it. It would stink to lose my instructor, but I could tell he was thrilled to be moving on to chase his dream. Who was I to get in the way of that?

"I'll set you up with another instructor," Jeff added, perhaps worrying that my congratulations weren't as heartfelt as they really were.

I ended up with an instructor new to the school and was glad for that. The most difficult part of completing training was having three things come together: availability of a plane, availability of an instructor, and good weather. As I mentioned earlier, I wasn't prepared for how difficult this could be. But a new instructor will often have a lot more availability since they haven't had time to build their book of business. If I could get in on the "ground floor," I'd have more of my new instructor's time to monopolize.

I met Connor on June 26. I had soloed on June 4, so to the extent there is a good time to make a change, this seemed fortuitous. One big milestone was behind me and another—my first solo cross-country flight—was looming.

Connor and I shook hands in the hallway outside the offices of East Coast Aero Club. Connor had an Irish last name to complement his Gaelic first name and he looked the part. He was about six feet tall with mischievous eyes and a youthful smile—a sort of Cheshire Cat grin that made you think he knew something you didn't, which isn't necessarily a bad thing if you're hoping to learn something from the person wearing the grin.

To say I was surprised by his youth would be an understatement. Connor was just twenty-one. He was younger than my own children. He was born *fifteen years* after the release of *Top Gun's* first installment. When he shook my hand and introduced himself to me, his shy nature made him seem even younger. He had gotten his pilot's license while an undergraduate student at a school in the western United States. His CFI certification came shortly thereafter. Like Jeff, he was working his way toward the airlines and I, along with other students, was a willing accomplice in his effort to build time.

Any shyness Connor had when in the offices of the flight school disappeared when we stepped on the tarmac or hopped in a plane. He became a field general. While doing our preflight, he'd point to something and say, "What's this?" in a staccato burst. I'd answer. "What's it do?" he'd follow up. If I answered correctly, he just say, "Cool," but in a way that made me feel like my response had been decidedly uncool.

One day, when doing this grilling, he pointed to something in the engine and said, "What's that do?" I looked. I had no idea.

"I don't know," I said after several unfruitful seconds of staring at connections into and out of whatever we were looking at.

"Yeah, I don't know either," he said, deadpan. "I was hoping you knew because I have no idea."

I laughed. He didn't.

Our first flight together we did pattern work. I was nervous. I had the distinct sense that Connor wouldn't just be judging me as we flew together but would also be judging my last instructor. Did Jeff train me well? Did I know what I was doing? For some reason, this nervousness weighed on me more than if I thought it was just my own reputation at risk. But I acquitted myself well enough on my first flight with Connor, and under difficult circumstances. In my logbook, he wrote the following:

Normal T.O. & Landing, Crosswind T.O. & Landing,
Go Around, Winds 15-G25!

For most of July and August, Connor and I practiced standard maneuvers. The entries from my logbook during this time were stuffed with repetitive text in the "remarks" column of my entries: slow flight, turns around a point, stalls, go-arounds, short-field takeoffs, soft-field takeoffs, and more. We were hitting most, if not all, of the maneuvers listed in the *ACS*.

As I became more proficient and comfortable in the airplane, I began to allow myself to think about the day I would have my license. And that led me to think about flying with my wife, Mary. She's been a nervous flyer her entire life and, like most people with a fear of flying, her nervousness came from experiences on large, commercial airliners. I knew if turbulence bothered her in a 150,000-pound 737 with all sorts of redundant systems, then it was likely to bother her much more in a 2,000-pound single-engine, piston-driven airplane.

This concerned me. Would she be willing to get into the cockpit with me when I eventually did get my ticket? And if not, what was I doing this for? While flying alone or with a buddy could be fun, sharing adventures with Mary was much more appealing.

As I considered this question, I thought about something I'd noticed throughout my life: those we love and who love us don't tend to trust us with complex tasks as they might a stranger. I knew that my wife, at least early on in my pilot journey, would always think somebody she didn't know was more competent than me. This realization led to an idea: I decided to hire my new, young instructor to take both of us up together.

This is something I'd recommend if you hope to fly with a friend or family member who is likely to be a nervous passenger. It's a great way to get somebody to understand what general aviation flying is like before they experience it with you alone. You don't want to be dealing with a freaked-out passenger, particularly as a low-time pilot. Anything that takes you away from focus on the task at hand—flying—is a risk. And trust me, a passenger in a disrupted mental state is a huge distraction.

On the day of our scheduled flight, and after introducing Mary to Connor, we made our way out to the ramp and began preflighting the airplane. Connor did his usual routine. He rapid-fired questions at me and, with each correct answer, said, "Cool." I had the distinct impression Connor was relishing the opportunity to demonstrate his teaching method in front of my wife. That suspicion wasn't made any less acute when, later that day, Mary would comment that she rarely had seen me flustered in all our time together, but she did during that preflight. I think the authority with which he dealt with me gave Mary confidence that he knew what he was doing. I also think she loved seeing me humbled by such a young instructor.

The flight went well when it came to introducing my wife to the magic of general aviation. The day was gorgeous. Visibility was unlimited. Boston came into view quickly after we climbed out of Hanscom, a gleaming, futuristic image with the Atlantic Ocean as a backdrop, reflecting bright sunshine. I heard Mary say breathlessly and repeatedly, "Oh my God…." This was in direct violation of my "sterile cockpit" rule which I had briefed her on before we took off.[50] It later occurred to me that I had neglected to tell her that even though she wasn't talking to us and was speaking in very hushed tones, the intercom made anything she said audible to us.

I had my hands on the controls throughout the flight. After doing some sightseeing, I did a touch-and-go at an airport about 20 miles from Hanscom and then we returned.

"What did you think?" I asked Mary.

"I think we should hire Connor to fly with us after you get your license." I laughed. I wondered if my plan had backfired.

Navigation

Flying with Mary and Connor gave me a taste of what I looked forward to: going somewhere with my wife once I had my ticket. I'll bet that's why you are getting or have gotten your license too; you also want to go somewhere. You're not going through all this pain just so you can fly a couple of miles from your home airport and do the maneuvers listed in the *ACS*. You want to go from your home airport to one far away and visit relatives or spend the night in a bed and breakfast. If you want to feel the romance of flight, you'll have to navigate between locations.

---

[50] The sterile cockpit rule prohibits conversations unrelated to flying the aircraft during critical phases of flight such as takeoff and landing.

"In-dash" GPS systems and EFBs like ForeFlight, have made getting from point A to point B as easy as "flying the magenta line."[51] But you won't be allowed to navigate this way during much of your training. You'll be taught to fly the old-fashioned way: by dead reckoning and pilotage. Dead reckoning is defined by the FAA as "navigation solely by means of computations based on time, airspeed, distance, and direction." Pilotage is navigation by reference to landmarks and checkpoints. When I say these methods are old-fashioned, I mean *really* old-fashioned, particularly dead reckoning. Dead reckoning existed in the nautical world before it was used in the aeronautical world. Ship's captains would use their estimated speed through the water and heading, along with celestial navigation, to keep track of their position on the ocean. Evidence of using dead reckoning to find your way from one place to another exists at least as far back as the thirteenth century.

I had learned through my Sporty's online course and various free online videos how to use a VFR navigation log to plan for a cross-country flight. A cross-country flight is often mistakenly defined as flying to an airport more than 50 nautical miles from your point of departure. It *is* true that you need to fly to a destination more than 50 miles from your departure point for the purpose of "meeting the aeronautical experience requirements … for a private pilot certificate." But outside of this phase of your pilot journey, a cross-country flight is simply one that includes "landing at a point other than the point of departure" and that "involves the use of dead reckoning, pilotage, electronic navigation aids, or other navigation systems to navigate to the

---

[51] The line showing the path between two points in a GPS flight plan is a magenta color.

landing point." The bottom line: if you land somewhere other than where you took off from, you probably completed a cross-country flight.

You'll become very familiar with the VFR navigation log during your training and will look forward to the day when you won't have fill one out again unless, of course, you are an E6B master and relish the use of the antiquated slide rule to prove your pilot bona fides. Even for the practiced student pilot, filling out the spreadsheet-like log is tedious, especially for a long trip with many checkpoints.

Speaking of spreadsheets, I'd recommend finding or creating a VFR log in Microsoft Excel or some other spreadsheet tool. If you have a knack for writing simple formulas, you can automate some of the calculations in a way that makes it easy to adjust your numbers if, for example, the wind speed or direction changes between your first go at a given flight plan and one closer to your departure. It also allows you to save a route so that you can retain your checkpoint sequence and just update the numbers in the rest of the grid if you plan to fly the route more than once.

Given how tedious completing the navigation log is, I won't spend time here giving step-by-step directions on how to complete every single cell in the grid (although I will cover some of the more critical calculations in a moment). There are plenty of online resources to help you with this. In fact, my own Google search returned 693,000 results, of which 29,300 are videos. I found these resources very helpful while learning about navigation, and I'm sure you will too.

You are probably waiting for me to be critical of the requirement that student pilots use this manual, anachronistic process to plan a trip when computer-driven tools are so much easier to use—not to mention more accurate. After all, if the

E6B, which I lambasted earlier, is the tool of choice for many when doing calculations to fill out the navigation log, then isn't the log itself equally useless?

No.

Here's why: filling out the log forces you to immerse yourself in several concepts that are important if you are going to be a complete and safe pilot.

The sectional map

The first major benefit of filling out the log is that it forces you to look at a sectional map and really understand it.

Aviation charts are probably the oldest of aviation tools. They first began to appear in the 1930s as air travel became exponentially more prevalent. The history of this navigation aid is captured nicely—and very visually—by the United States Library of Congress. The world's largest library has a great online collection of very old aviation charts. I found one that depicts the area around my parents' hometown of Albuquerque, New Mexico. A few things struck me about the chart. First, it lists the population of the "city" as of publication of the chart at 25,000. Albuquerque is now over half a million in population. Secondly, the chart doesn't look much like a modern sectional. It just looks like … well … a map. And a not-too-busy one at that. There are no airspace designations, VOR compass rose overlays, radio frequencies, or military operations areas. In fact, the only indication that it isn't simply a terrestrial navigation map is the linear layout of icons depicting the location of lighted beacons pilots used to navigate in an era before more sophisticated navigation tools.

The airway beacon system was constructed when all flights were done by visual navigation; there were no tools for instrument flying. A *U.S. Centennial of Flight Commission* article noted that in 1919 an Army Air Service lieutenant named Donald Bruner used bonfires across the countryside to help aid pilots flying at night find their way from one city to another. In 1921, a pilot successfully tested this system by flying at night from North Platte, Nebraska, to Chicago. The success of bonfires as navigation aids led to a more formal system of lighted beacon towers. The U.S. Postal Service worked to complete the system, which facilitated faster and safer airmail deliveries. By 1933, 18,000 miles of lighted airways were complete.

On the old Albuquerque aviation map, I noted that next to the icons designating the location of the lighted beacons, there was what appeared to be a Morse code designator. At first, I had assumed this was some sort of audible radio identification for each beacon. But it turns out it was a visual code—the lighted beacons would emit long or short bursts of light. The longer bursts were a "dash," and the shorter ones were a "dot." The towers were arranged in a sequence where each Morse identifier was one of ten letters: W, U, V, H, R, K, D, B, G, and M. Pilots would remember the expected sequence of identifiers by way of a mnemonic: "When Undertaking Very Hard Routes, Keep Direction By Good Methods." Sure enough, the Albuquerque map follows this sequence.[52] Thus began the use of word tricks by pilots to remember important things.

Today's sectional chart that includes the Albuquerque area looks much different. It has the large Class C airspace over KABQ. It has Class E airspace carved out of the sky around

---

52 Some of the structures that housed these old beacons are still standing. In Milan, New Mexico, the Aviation Heritage Museum has restored a tower and support buildings. Others exist around the rest of the country too.

airports sprinkled through the water-starved, brown land. It depicts the locations of VORs and victor airways. Fortunately, this busier chart is indicative of a more modern and safe air system. (I'll go into the details of airspace in the next chapter.)

While you won't have lighted beacons to reference, one of the first steps when you fill out your cross-country log is to identify checkpoints (some say "waypoints") along your route. This will require drawing lines on your map between points on the ground that you are certain you can see from the air. When doing this, I tried to keep the distance between the two points to no more than 15 to 20 nautical miles. This ensures that you have checkpoints appearing frequently enough to keep you on course. If you are off course a few degrees when navigating between two points relatively close together, it won't be a big deal as the visual reference you are looking for on the ground should always be in your field of view even considering such an error.

As you draw your lines, be sure to consider airspace restrictions (which we'll cover in the next chapter) and terrain altitude. Just because you can draw a line doesn't mean you can fly it. Also note that many students aren't great—me included—at picking ground references that are easily visible from the air. After gaining some experience during cross-country flights, I've come to note things that were more visible than others. With those lessons in mind, I'd recommend the following checkpoints:

- Airports—depending on size, these tend to be visible with relative ease. Also, you'll want to have emergency landing locations along your route in any case, so flying near or over airports has an added safety bonus.

- Road intersections—assuming the roads are large (like freeways) and intersections are not too frequent (many intersections in a small area make it hard to tell which one you are looking at), intersections can be visible from very high altitudes.
- Stadiums—large stadiums are often clearly visible, particularly when they have expansive parking lots around them. (An important note—stadiums where professional teams play will often have TFRs over them. A TFR is a temporary flight restriction which does exactly what it sounds like: prohibit aircraft from entering the airspace over a location for a defined period. If you plan to fly over a stadium, be sure no TFRs are active.)
- Distinctive bodies of water—I say "distinctive" because in some areas, ponds and lakes are abundant. This is the case where I fly and in instances where I chose one of them, I found that it was not very easy (and often impossible) to find the one that looked so obvious while doing my preflight planning. If the body of water is exceptionally large or is by itself in a large area, it will stand out well.
- Large bridges—rivers of moderate to large size are easy to see. Large bridges crossing them can be as well. If you see a road crossing a river on your sectional, you should see a bridge.
- Distinctive prominences—land features that stand out from topography of the area where they're located are wonderful navigation aids. For example, a large hill or mountain that suddenly springs up out of flat land works well. I want to emphasize the word

"distinctive" again though. An individual mountain peak, for example, is hard to identify from the air when it's among a chain of other mountain peaks.

- Towns—even small towns are quite visible from altitude. But it's very difficult to know for sure you're looking at the town you had in mind if it's one among many in a small area. As with other features mentioned in this list, the more isolated the town is, the easier it will be to know you're over it.

Speaking of being "over" something, don't worry about flying directly over a checkpoint. In fact, I'd suggest you draw your course line so that the checkpoint is off to one side of the airplane, preferably the left side. The nose of your aircraft will make it difficult to see what's ahead of you for many miles. And obviously, once whatever you're aiming for passes under the nose, you won't be able to locate it.

Also note that your eye is tuned well to notice linearity out of chaos. Things that have long lines to them—roads, rivers, and railroads—tend to stand out in all terrain. While they are great references if you're going to fly along them as if you were a boat on the river or a car on the road, they aren't very useful as defined waypoints *unless* you have a second identifier coupled with them. For example, a smokestack along a river is a great waypoint. But drawing a line over a river on your sectional where there isn't another prominent feature will mean you won't know for sure if you're crossing where you expected to.

Course and Heading

After you've drawn your lines on your sectional between waypoints, you'll fill out your VFR log. When doing so, you'll quickly come to understand the difference between course and heading. That's because you'll spend a bit of time translating the neatness of the plan your draw on your sectional into information that accounts for the messiness of the natural world. In fact, that's what the columns about heading and course are all about—figuring out where to point your airplane based on what your compass (your heading) says so that you can fly the path over the ground represented by the lines you've drawn on your sectional chart (your course).

To start your navigation planning, you will begin by measuring the angle between the course you desire (the one you drew on the map between checkpoints) and a line of longitude, one of the north/south lines on the map. For example, if your first checkpoint is due east of your departure airport, the angle between the line you drew on your sectional and a line of longitude will be 90°. You will write this number into the "true course" cell:

True Course	True Heading	Magnetic Heading	Final On Cousre Heading
+/-Wind Adjustment	+/- Magnetic Variation Adjustment	+/- Magnetic Deviation Adjustment	
90°			

Your true course describes the path you want to trace on the ground in a world with perfect conditions. It's called a true course because it's your course in reference to true north—the geographic north pole. But there are two things that make flying true course to get to where you want to go problematic. The first is wind. If you just flew the 90° in our example, you wouldn't end up where you intended if there was any wind at all—and there is *always* some wind. The good news is that accounting for wind is easy. After looking up what the winds are predicted to be at your planned cruise altitude, you can use your E6B or an electronic flight computer to enter the wind speed, wind direction, the course you want to fly, and your true airspeed (more on this later) and, voila, you'll have your wind correction angle.[53]

For example, assume I did indeed need to fly a true course of 90° and planned to do it at a true airspeed of 110 knots when the wind was from 45° at 15 knots. Using my electronic flight computer, I can quickly determine that my wind correction angle is -6°, meaning I must subtract 6° from my true course of 90° and instead fly a heading (we call this "true heading") of 84° to stay on track. This makes sense intuitively. If I want to fly due east at 90°, and a wind is coming from 45°, that means I am being pushed from the front left—from about my 10 to 11 o'clock position. Therefore, to ensure that the path I trace on the ground is truly 90° as it appears on my sectional, I must point my nose to the left to correct for this, and 84° is to the left of 90°. In our log, we write the wind correction angle of -6° under our true course. We then write in the true heading cell in the next column over the result of 90°-6°, or 84°.

---

[53] Note that wind directions in printed forecasts are referenced against true north—i.e., the angle the wind is blowing relative to a longitudinal line, *not* magnetic north, which will be explained shortly. In fact, a well-worn phrase in aviation is, "If you read it, it's true; if you hear it, it's magnetic." When you hear a wind direction from the tower or on an ATIS or ASOS, it's given in reference to magnetic north. This makes sense; since runway headings are referenced to magnetic north, it helps you instantly understand where the wind is coming from relative to the centerline of the runway and your compass heading.

True Course +/-Wind Adjustment	True Heading +/- Magnetic Variation Adjustment	Magnetic Heading +/- Magnetic Deviation Adjustment	Final On Cousre Heading
90°	84°		
-6°			

Your true heading, then, is your true course adjusted for wind drift. But now we need to account for the other curveball nature throws our way. We called the course we wrote down the "true course" because it's a course relative to true north. True north points to the geographic north pole of the Earth, the place where all lines of longitude converge on the globe. Unfortunately, though, the geographic north pole is not where the compass in your airplane points. It points to the *magnetic* north pole. And as you've probably surmised by now, the magnetic north pole is not in the same place as the geographic north pole. The fields of force that tell your compass where to point don't concentrate themselves at the physical north pole where longitude lines deceive you into believing they should. In fact, they don't really concentrate themselves anywhere for long. They are continually moving. As of the writing of this book, the north magnetic pole was about 600 miles from the physical north pole. In 1859, it was positioned about 1,400 miles from the north pole in extreme northern Canada. We call this disparity between where our magnetic compass points and where the geographic north pole is "magnetic variation" (or "magnetic declination").

This is the second adjustment we need to make to our sectional map world of true north and true course—an imaginary world that doesn't really exist when it comes to using your compass to navigate. You must account for magnetic variation. Fortunately, this too is easy to do. On your sectional, you will see periodically spaced, dashed lines running diagonally across the whole of the map from top to bottom. These lines are called isogonic lines and they tell you what the magnetic variation is in the area where you'll be flying. In your navigation log, underneath your true heading, you will use this information to enter an adjustment. In my neck of the woods, Massachusetts, as of this writing, the isogonic line is labeled "13° West." When the declination is a westerly one, we add it to our true heading. When it's easterly, we subtract it. This navigation rule may remind you of a mnemonic you've probably heard, "East is least, and west is best." Since my isogonic line tells me the magnetic variation is "west," I add the number of degrees on the line to my true heading: 84° + 13° = 97°. This is my magnetic heading; I write this number in the cell labeled as such. *This* is what my compass should read given wind and magnetic variation if I want to fly to my checkpoint which, on a sectional map, looks like it's exactly 90° from me.

True Course	True Heading	Magnetic Heading	Final On Cousre Heading
+/-Wind Adjustment	+/- Magnetic Variation Adjustment	+/- Magnetic Deviation Adjustment	
90°	84°	97°	
-6°	+13°		

Okay, I misled you a bit. As you can see above, there is one more cell where we need to enter some information—an adjustment due to magnetic deviation. Magnetic deviation is an error introduced to your compass reading due to magnetic fields in the airplane itself. The amount of deviation for a given heading is particular to your aircraft. Usually, you'll see a card in your airplane under your compass that shows the actual direction you should steer for a desired magnetic heading. It might look something like this:

For	N	30°	60°	E	120°	150°
Steer	359°	31°	60°	89°	121°	149°
For	S	210°	240°	W	300°	330°
Steer	179°	211°	239°	W	299°	331°

Note that the deviation errors are typically small. One to two degrees is the norm in most aircraft. For this reason, I usually left this correction out of my calculation. Since knowing with precision the actual wind speed or the exact value of magnetic variation at your precise location is nearly impossible, all of your calculations are guesses in any case, good guesses to be sure. But one or two degrees over 15 to 20 miles is not going to make that much of a difference. Besides, there's a more practical reason I left it off and you might to: if you bounce between airplanes a lot during your training, you probably won't know in advance what the deviation card in each aircraft reads. So don't stress if you don't have a number to put into your log to make one last correction to your magnetic heading. It really won't matter.

True Course +/-Wind Adjustment	True Heading +/- Magnetic Variation Adjustment	Magnetic Heading +/- Magnetic Deviation Adjustment	Final On Cousre Heading
90°	84°	97°	
-6°	+13°	0° →	97°

When you're done with all these calculations, you can put the number you end up with, 97° in our example, in the "Course" or "Route" column of your log, which is usually in a cell just to the right and between the relevant checkpoints. This number is what you'll want to see on your compass if you hope to make it to the next checkpoint on your route.

Check Points	On Course Heading
Home Airport	97
Stadium	

Airspeed

Knowing the direction to fly is only part of the navigation puzzle in a dead reckoning world. Recall that the definition of dead reckoning according to the FAA is "navigation solely by means of computations based on time, airspeed, distance, and direction." We've figured out the direction part. Now we need to figure out the time, airspeed, and distance part of the puzzle. These aren't really three different things to consider though. You may recall from high school this formula: $d=rt$. The "d" is for distance, the "r" is for rate (otherwise known as speed), and the "t" is for time—"distance equals rate times time." If you know two of the things in this formula, you can figure out the third.

What we care most about calculating is the time to fly between our checkpoints. We want the time for two reasons. First the sum of the times between checkpoints tells us how long our flight will be—something most of us want to know when planning a trip. Second, we want to know, once we point our plane in the direction our earlier calculations told us to, when to expect to look down and see our next visual checkpoint, ensuring our navigation calculations were correct. To calculate time, we manipulate the $d=rt$ formula into this: $t=d/r$. If we have distance and rate, we know time.

The good news is that we already know the distance part of the equation. We can measure this in nautical miles on our sectional chart with a scale, a ruler-like tool with tick marks denoting distance.[54] You just place the scale along the line between your two checkpoints on a sectional map and count the marks. Let's assume for the purposes of creating an example that we are flying between a smokestack and a stadium, and the distance is 18 miles.

You may now begin to get excited. We have our distance. And we've already decided on the speed we want to fly earlier in our discussion—110 knots. Knots is another name for "nautical miles per hour," so we even have the distance portion of our speed in the same units as our distance on the ground, making the math straightforward. So, can't we now calculate the time to fly between checkpoints? If it were only so easy….

We did indeed decide on a speed. And we have an airspeed indicator in the cockpit that tells us how fast we are going. But that airspeed indicator tells us how fast we are going through air of a particular density, not over the ground (it's why we call it an *air*speed indicator!). And what we care about is our speed over the ground because this is what dictates how fast we cover the distance between two checkpoints.

---

[54] This is as good a place as any to explain the difference between a nautical mile and a statute mile (sm). A statute mile is used in the U.S. as a standard measure of distance on things like road signs telling how far it is to the next rest area. At 5,280 feet, it has a long history beginning with the Romans in the ancient era. A mile for them was 1,000 paces where one pace was every other step. In fact, the word "mile" derives from the Latin *mille* which means 1,000. This original definition of a mile morphed over time to mean the standard 5,280 feet we know today. A nautical mile came later and was defined as 1/60 of the 360 degrees that define the circumference of the Earth. This isn't a perfect way to measure a nautical mile though because the Earth isn't a perfect sphere. Today, the nautical mile is 1/60 of a degree of a hypothetical sphere approximating the size of the Earth and is fixed at 6,076 feet (1,852 meters).

Remember when I talked about airspeed earlier in the book, I noted that your airspeed indicator measures the difference between the pressure of moving air hitting the airplane (called ram air) and the static air pressure. This may seem like an unnecessarily complicated way to measure speed at first blush. But doing so helps us in a huge way. It allows us to know, *regardless of air density variations* due to altitude, humidity, temperature, or other factors, that enough air is getting over the wing to keep the airplane flying. Said another way, things like the stall speed of the airplane and rotation speed for takeoff don't change because of air density. If your stall speed in a landing configuration at a certain weight is 48 knots indicated airspeed, it's 48 KIAS whether you are at 10 feet or 10,000 feet.

Imagine for a moment that instead of having a speed indicator that accounted for air density you had one that just told you your ground speed (something you do have in the cockpit if you have GPS). Think of how difficult it would be to fly an airplane with only this measure of speed. Taking off in Boston (essentially at sea level) on a calm day you might need a ground speed of 55 knots to get airborne. But in mile-high Denver, you might need 65 knots. If you were taking off into a stiff wind in Boston, you might be able to get airborne with just 40 knots of ground speed. With a tailwind? You might need 60 knots. The pitot/static system that drives your airspeed indicator gets rid of this confusion. It doesn't matter if you are in Boston or Denver. It doesn't matter if it's hot or cold. It doesn't matter if you have a headwind or a tailwind. If the takeoff speed (as shown on your airspeed indicator) for your airplane at a given weight is 55 knots, it will be 55 knots no matter the air density or wind direction. If the stall speed in a landing configuration is 45 knots, it will be 45 knots all the time. The airspeed indicator, in fact, isn't really a speed indicator—at least in the way we traditionally think of speed—at all. It's a dial that tells you the condition of air flowing over the wing. And that is all that really matters to a pilot.

It's all that really matters, that is, until the one time you want to know your ground speed—when filling out your navigation log! To bridge the gap between what you chose as your cruise speed and what that means for your ground speed, you'll need to know a few pieces of information.

The first thing you'll need to get is your plane's calibrated airspeed. Calibrated airspeed (KCAS—Knots Calibrated Airspeed) is KIAS corrected for position and instrument errors. Getting into the weeds of KCAS is too much for this author and this book but suffice it to say that because it's hard to get a single instrument to measure airspeed perfectly in all circumstances, engineers create a table that shows what the airspeed would be if you did have such perfection. You can find this table in your POH and can use it to convert from KIAS to KCAS for a given flight condition (like cruise flight). For example, if you plan to fly at 110 KIAS in a Cessna 172S while at cruise, your KCAS according to the table in the POH is 105 knots.

Now that you have your KCAS, you use this to calculate your true airspeed (TAS). True airspeed is a measure of how fast your aircraft is zipping by air molecules just outside your window. Remember that our indicated airspeed doesn't measure this. An airplane flying at 2,000 feet and 110 knots is flying by adjacent air at a slower speed than when it is flying at 10,000 feet at 110 knots. This is because as the air thins, if the plane is to maintain that 110 knots indicated airspeed, it needs to move through the thinner air faster to "ram" enough molecules into the pitot tube to maintain a given ram air pressure that correlates to 110 knots. In fact, true airspeed increases about 2% for every 1,000 feet in elevation gain (although this is just an approximation as it varies with temperature and pressure).

Given the relationship between TAS and air density, it's not surprising that calculating TAS requires you to input into your flight computer two air density factors: pressure altitude and temperature. We'll spend a good bit of time on pressure altitude later but it's simply what your altimeter reads if you put 29.92 inches of mercury into its Kollsman window.[55] This reading, along with calibrated airspeed are enough for your computer to spit out TAS. If the pressure altitude at my cruise altitude is 3,5000 feet, the air temperature at that altitude is predicted to be 5° C, and my KCAS at cruise is 105 knots, then my estimated true airspeed is 110 knots. If my predicted pressure altitude is 10,500 feet (I'm now flying through thinner air), the outside air temperature is -9 C, and I still plan to fly at 110 KIAS (so my KCAS is still 105 knots), then my TAS is 122.3 knots.

Phew. That's a lot of numbers. By way of reminder, we are working our way toward calculating our ground speed. The good news: we now have all we need to calculate it.

Whatever flight computer you use, you'll now feed it the forecast wind direction, wind speed, your true course, and TAS to get to your ground speed. In our example from earlier, we assumed that the wind speed was 15 knots from 45° and that we were on a 90° true course. At a true airspeed of 110 knots, our ground speed will be about 99 knots.

Now, we return to our formula for time: *t=d/r*. If our distance is 18 nautical miles, then our formula is t=18nm/99

---

[55] The Kollsman window is embedded in your altimeter. It contains a scale of pressures that you set using a nearby knob based on the local altimeter setting.

knots. Plug those into your calculator and 0.18 will show up in your display. Keep in mind this means 0.18 *hours* (since our 99 knots is the same thing as 99 nautical miles per *hour*), meaning the time between checkpoints is 10.8 minutes or 10 minutes and 48 seconds. These numbers are too precise; it's unlikely you will hit them exactly given all the accumulating, small errors in every measurement and calculation we make. But it's fair to say that sometime between 10 and 11 minutes after starting your timer over the smokestack, you'll be passing over the stadium.

Time to climb and time to descend

The example I gave in the previous many pages assumed you were already at cruise flight. But obviously, part of your flight will be climbing from your airport to your cruise altitude and descending back to an airport from that same altitude. And you will do both at speeds other than the one you chose for your cruise. When you climb, your speed will almost always be less than your cruise.

Your plane's POH will have graphs or tables that allow you to estimate your time to climb, fuel consumed, and distance covered. I say "estimate" because I find most of the graphs provided in the POHs are crude, to say the least. Early on in my training, I got hung up on trying to get a precise number out of those graphs. Don't do the same. An estimate is good enough. Remember, there will be accumulating errors throughout your cross-country planning. But your log allows you to understand these errors as you fly. To the far right of the log there is a place to record your estimates with respect to time between checkpoints and ground speed. You also will have places to record your actual time and calculate your actual ground speed. Doing this will help you to know if you are ahead or behind schedule, and thus whether you should anticipate reaching your next checkpoint earlier or later.

Chapter 16
Airspace

Mastering cross-country flights will also require you to become a master of airspace. And, if you are like most student pilots, one of the things that will perplex you initially is the way the FAA has carved up pieces of the sky into varying types of airspace. At first glance, things seem logical and simple. There's Class A, Class B, Class C, Class D, and Class E. So far so good. If you were given this progression on an I.Q. test and I asked you what the next letter in the sequence would be, you'd confidently say, "Class F," and you'd (sort of) be wrong. In the U.S., you skip straight to Class G airspace. I say "in the U.S." because the International Civil Aviation Organization (ICAO) does indeed define Class F airspace, but the FAA doesn't use it. And thus, you are introduced to the first confusing thing about airspace if you are flying in the United States.

Okay, so there's A, B, C, D, E, and G. What's the difference between them? First thing's first. The only "uncontrolled" airspace among Class A, B, C, D, E, and G classifications is the last one: G. Uncontrolled means that ATC "has no authority or responsibility to control air traffic." This may seem straightforward, but it's not for reasons I'll explain later. For the moment, let's focus on the controlled airspace, which is simply defined as airspace in which air traffic control can provide services to aircraft.

Class A
You aren't going there (not yet)

We'll start with Class A, and not only because it's airspace defined by the first letter in our not-so-obvious progression of letters. No, let's start there because, to my mind, it's the easiest to understand. And that's because your private pilot certificate, by itself, won't allow you to fly in Class A airspace. That's not because of the altitude levels that define the airspace: from 18,000 MSL up to 60,000 feet MSL. You certainly can fly an airplane with your PPL that has the capability to get to 18,000 feet MSL or higher (and has oxygen). But if you are operating that aircraft in Class A airspace, you must be on an instrument flight plan. And since you won't have an instrument rating when you complete your training and take your practical exam, you won't be up at those nosebleed altitudes. You will be tootling around in the air beneath Class A airspace.

Class B
The BIG airports

Class B, C, and D airspace are more relevant in your PPL world and are the next easiest to tackle. They are the airspaces around many airports, but certainly not the majority. Not even close. Depending on the data you look at, there are somewhere on the order of 5,000 public use airports in the United States and—get this—about 15,000 private use airports.[56] There are just under 600 airports with Class B, C, or D airspace around them. But note that the distribution of airports across these three airspace classifications is not equal. In the United States, fewer than 40 airports have Class B airspace around them. About three times as many sit under Class C airspace. And ten times as many airports sit in Class D airspace as sit in Class B.

I described airports as "sitting under" an airspace. This obviously isn't by accident. The airspace is there *because* of the airport. As you might expect from the breakdown in the number of airports with a B, C, or D classification, Class B airspace surrounds the country's busiest airports and is therefore the most restrictive; Boston Logan, Los Angeles International, and Atlanta's Hartsfield-Jackson are Class B facilities. You've probably already heard this airspace described as an "upside-down wedding cake." This is a good way to visualize Class B airspace. If you are inside the wedding cake, you are in the "Bravo" airspace.

Understanding the "why" behind the shape of Class B airspace is not hard and will help you understand better why the air is carved up the way it is in other contexts. Let's look at Boston Logan's (KBOS) Class B airspace.[57] (If you don't live where I live, you may not have a sectional of the area, so jump online to find a digital sectional.) The Class B airspace is

---

[56] A private use airport is not defined as such because of its ownership but because it requires prior approval for its use by an aviator. A public use airport does not require prior permission. Also note that our definition of airport here includes heliports and other such aviation-related areas.

[57] The *K* identifier at the beginning of the airport code is for airports in the contiguous United States. The identifier for airports in Alaska and Hawaii is *P*. In the United Kingdom, it is *E*.

identifiable on the map as a series of concentric, solid blue rings around Boston Logan Airport. The inner ring of the Boston airspace shows up on the sectional as a blue circle about fifteen miles in diameter. Inside that ring, you'll see the following: $\frac{70}{\text{SFC}}$. This means that that inner ring defines a cylinder that goes from the surface to 7,000 feet MSL (add two zeros to the numbers that appear in these "fractions" describing layers of our cake to get the altitude in feet MSL). If you are between the surface of the Earth and 7,000 feet MSL and within 7.5 miles of the airport (the radius of our cylinder), then you are in the inner ring of the Boston Class B airspace.

The next ring you can see on the sectional is larger in diameter, about 20 miles. This ring has the following altitude descriptor: $\frac{70}{20}$. This cylinder, then, is described as 20 miles in diameter beginning at 2,000 feet MSL and rising to 7,000 feet MSL. One thing you'll notice is that both cylinders have the same "top" altitude. This altitude defines the very top of the airspace or, if you want to look at it another way, the bottom of our wedding cake before we turned it upside-down.

The outer ring of the Boston Class B airspace is about 40 miles in diameter. It shows a top—surprise, surprise—of 7,000 feet MSL and a bottom of 3,000 feet MSL (with the exception of a piece of the airspace to the west which is 4,000 feet MSL).

The reason the airspace descends closer to the ground as you get closer to the airport is probably obvious. As you near the airport, aircraft are going to be operating at lower and lower altitudes as they take off and land. Air traffic will become denser as well. Therefore, to ensure adequate separation between aircraft in an environment like this, you must be under control of ATC when in the airspace. In fact, thinking of this underlying purpose of the airspace helps you visualize why Class B airspace is structured the way it is. Imagine drawing lines representing the instrument departure and arrival procedures aircraft follow on their way into and out of a Class B airport. Now imagine building airspace around these lines for up to twenty miles from the airport. You would draw an upside-down wedding cake too.

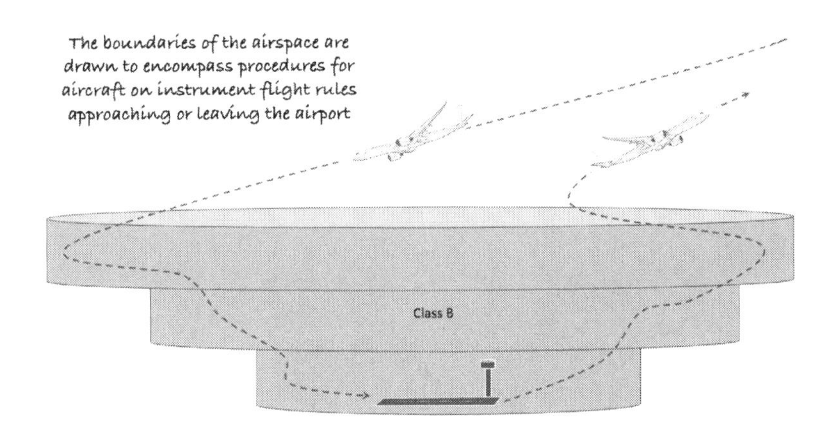

"No, I wouldn't," you might say. Why not, you might argue, make the airspace one giant cylinder? One that is the diameter of the outer ring (40 miles) and runs from the surface to 7,000 feet MSL? This would be easier, but it would make a large portion of the sky around big cities inaccessible to a lot of general aviation traffic. Yes, it's true that you can be cleared into Bravo airspace as a general aviation pilot. But it's never guaranteed and, at least in my experience in the Boston area, it's rare. I think I've been cleared into the Bravo about 20% of the time when the shortest path to my destination would have taken me through it.

By not gobbling up all the airspace, the regulators have left some room for those of us not flying commercial airliners and other aircraft going in and out of Class Bravo airports. In fact, if you look more closely at the Boston area sectional, you'll note a lot of smaller airports that sit under the Bravo, including my home airport, Hanscom Field (KBED), to the northwest of Boston. If the Bravo airspace were structured as a single, giant cylinder, the moment you left the ground at Hanscom, you'd be in Bravo airspace. This wouldn't only be a pain for you as a pilot, but it would also be a pain for the controllers of the Bravo—imagine having to coordinate every single aircraft taking off from Hanscom and any of the other five airports within that twenty-mile radius of Boston Logan.

The regulators recognized the unnecessary workload this would create and drew Bravo airspace to be as accommodating as possible to other traffic. To that point, if you look closely at the Bravo airspace around Boston, the northern and western part of the outer shelf—from about the 7 o'clock position to the 12 o'clock position—is modified. While most of the bottom of the shelf starts at 3,000 feet MSL, this part starts a 4,000 feet MSL. The airspace was drawn this way because the orientations of the runways at Logan don't necessitate bringing traffic into Boston from that direction at lower altitudes; none of the flight paths dictated by approach or departure procedures are below 4,000 feet MSL in that sector. Realizing this, the regulators decided to give a part of the sky back to aviators who might need to stay outside of the Bravo.

In fact, the guidance the FAA provides in documentation that addresses the design of Class B airspace gives regulators surprising latitude to exercise good judgment in the airspace design (from Order JO 7400.2N):

> *There is no standard Class B design. Instead, the size and shape of the Class B airspace area will vary depending upon location-specific ATC operational and safety requirements. The Class B airspace design should be as simple as practical, with the number of*

sub-areas kept to a minimum. Its vertical and lateral limits must be designed to contain all instrument procedures at the primary airport(s) within Class B airspace.

See … the FAA has your interests in mind. To the extent possible, they want Class B airspace to be simple, and limited. But they aren't so accommodating as to let you go into the Bravo airspace whenever you see fit. As I mentioned earlier, I've been denied clearance into Bravo airspace more than I've been cleared into it. To get into the Bravo, unlike other airport airspaces, you must get very explicit permission; you must hear the magic words: *cleared into the Bravo airspace.* This is not true for Class C and D airspace.

Class C
The middle child of controlled airports

Class C airspace surrounds airports that are not quite as busy as Class B airports. Indianapolis, Indiana, Savannah, Georgia, and San Jose, California are all Class C airports. On a sectional map, Class C airports are identified with solid magenta rings around them. If you look at Lovell Field (KCHA), Tennessee, you'll see such rings.[58] A Class C airspace usually has two rings. At Chattanooga, they are $\frac{47}{\text{SFC}}$ and $\frac{47}{26}$. The inner ring is a cylinder 10 miles in diameter that goes from the surface to 4,700 feet

[58] By the way, if you're thinking Lovell field is named after Jim Lovell of Apollo 13 fame, you aren't alone. I assumed that it was named for him as well. But it's not. It's named after John Lovell, who was a civic leader in the Chattanooga area for decades, staying true to the axiom that most large public airports are named after political figures.

MSL; the outer ring is 20 miles in diameter and has a bottom of 2,600 feet MSL and a top the same as the inner ring: 4,700 feet MSL. Don't let the strange altitudes confuse you. Remember that these altitudes are MSL. Since the altitude of KCHA above sea level is nearly 700 feet, a 4,700-foot top to the airspace means that it's 4,000 feet above the ground, fairly standard for Class C airspace. The bottom of the outer shelf is around 2,000 feet AGL.

Class C airspace is more accommodating to general aviation aircraft. Not only is it usually easier to get clearance into Class C airspace but it's also easier to land at Class C airports than at Class B airports. In some cases, you need advance permission to land at a Class B airport and must be on an IFR (instrument flight rules) flight plan. Not so at a Class C airport. Unlike Bravo airspace clearance, permission to enter the Class C airspace is granted indirectly. If, after you contact the tower at a Class C airport, the controller responds with your tail number, then you may enter the airspace. For example, suppose I make initial contact with Lovell by saying this: "Lovell Tower, Cessna 123AB, 10 miles to the west with Foxtrot, full stop." If the tower responds, "Cessna 1234BA, Lovell Tower…" then I can enter the airspace. If, however, the response to my initial contact was "Aircraft calling Lovell Tower, standby," then I better be sure not to cross that imaginary magenta line in the sky until the tower gets back to me.

Another thing to be aware of when flying into Class C airspace is that these airports typically have regular departures and arrivals of commercial aircraft, although on a less frequent basis than Class B airports. This means you'll need to have your antennae up, not just for the purpose of keeping an eye out for traffic—something you should always be doing anyway—but because large, heavy aircraft will be flying, landing, and taking off nearby. This makes the risk of wake turbulence greater than you will find at Class D airports or nontowered fields. Wake turbulence refers to the moving air a large aircraft leaves in the atmosphere behind its path of travel. The air can be moving violently enough to easily flip a general aviation aircraft (and sometimes even larger planes). It's therefore important to leave significant distance (several miles) between your plane and a larger aircraft in front of you and to also fly above the path of the larger aircraft since wake turbulence "settles" over time. Again, I don't want to leave the impression that wake turbulence won't ever be an issue elsewhere. But the probability of encountering it in a Class C area is higher than in D, E, and G airspace.

Class D
A great place to learn

Many of you already in training fly out of a Class D airport. But if you haven't started your training and have a choice, I'd recommend flying out of a Class D airport if you can. Doing so gives you tons of exposure to using radios and flying in controlled airspace without all the hassles of a large Class C or Class B airport.

Class D airspace is defined by a single cylinder over the airport designated by a dashed blue circle on the sectional. There are no "shelves" to the airspace as you find at Class B and Class C airports. Examples of Class D airports include Martha's Vineyard (KMVY) near Cape Cod, Centennial Airport (KAPA) near Denver, and Palo Alto (KPAO) near San Francisco.

Class D airports are often not Class D at all times. Because air traffic at these airports tends to drop off at night (when a lot of general aviation traffic in particular decreases substantially), the tower isn't staffed 24/7. So, what happens when the control tower is closed? Usually, the airspace changes to Class E or G. The airport then becomes one where you use a frequency (usually the tower frequency) as a common traffic advisory frequency (CTAF) just as if you were at an airport without a tower.

Class E
The most confusing of airspaces

Okay, remember when I said that airspaces could be confusing? Well, we've covered the easy stuff. As we move up the alphabet, things become a bit trickier.

It's often said about Class E airspace that the *E* means "everywhere." And, indeed, it does seem to be everywhere; most of your flying time will be in Class E airspace (assuming you leave the traffic pattern every now and then!). The "everywhere-ness" of Class E is even made clear in the definition the FAA gives to the airspace in the *PHAK*: "Class E airspace is controlled airspace not classified as Class A, B, C, or D airspace." It's not often you see a definition that defines something by saying what it's not.

While the definition is a bit bizarre, it's not the main thing that new pilots get hung up on. That honor belongs to a single word in the definition: *controlled.* Class E airspace is considered *controlled* airspace. But why? After all, I'd venture that virtually every one of you has flown in Class E airspace without talking to ATC. And if you're not having to get permission to fly in an airspace or talk to anybody while you're in it, how is that airspace considered controlled?

The answer lies in the fact that you, as a pilot flying under VFR, are not the only person slicing through Class E air. You'll be sharing that air with pilots flying under IFR. And pilots on

IFR flight plans *are* controlled while in Class E. This makes sense; it would be hard to see and avoid other aircraft when you're in a cloud that happens to be in Class E airspace. You'll need ATC to help you with that. Class E, then, fills the gap between the Class A, B, C, and D airspaces so that planes on IFR flight plans *are* controlled and provided separation from other aircraft even when they're not near an airport or up in the Class A realm.

If you've flown a longer cross-country flight with flight following, you've probably gotten some insight into how the various FAA assets are linked together to provide control for IFR pilots. Flight following is just what it sounds like: asking the controller to set you up in the ATC system so that controllers can follow your flight. Setting up flight following (also known as "advisory services") entails having the controller provide you a unique transponder (or "squawk") code—a four-digit code that you simply type into your transponder. That code links you to an image on their screen that represents your aircraft and that includes information such as your location, pressure altitude, and ground speed. With your aircraft in the system, controllers can provide you with alerts about traffic that might be nearing you. But that's not all. In addition to traffic deconfliction services, I've also been alerted to skydiving occurring along my route and was advised to change my course to avoid the parachuters. I've been warned about areas of weather in my vicinity. I've even been the one providing the help before. I once was asked by a controller I was in contact with to confirm the location of a small forest fire.

Flight following is only provided, though, as controller workload permits. This is an important thing to remember. Their priority is IFR traffic. Even if you have been granted flight following (and I've never been denied, even in busy airspace), don't assume that you can fly with impunity when it comes to collision hazards. Twice while receiving the service, I came within a few hundred feet of another aircraft and the controller never said a word. In both cases, I was using an in-cockpit ADS-B traffic display and was made aware of the danger. Always assume that this could happen to you and, as they say, "keep your head on the swivel."

I once did a cross-country from my home airport of Hanscom Field (KBED) to Lebanon, New Hampshire (KLEB) and used flight following. Even though I was flying a VFR flight, I got a good sense of how the air traffic control network works for those on IFR flights. I started off my journey talking to the Hanscom Tower in the Class D airspace. Shortly after departing the field, I was handed off to Boston Departure and, shortly after that, I entered Class E airspace. A bit farther into my journey, still inside of Class E airspace, I was handed off to Boston Center. During my time tuned to Boston Center's and Boston Departure's frequencies, I could hear IFR traffic being given instructions by the controller. While I could have flown in that airspace without talking to ATC—flight following is optional after all—it drove home the point that while in Class E airspace, IFR pilots were still being controlled.

Understanding why Class E is controlled airspace helps us to understand the altitudes that define it. The "standard" vertical dimension of Class E airspace is from 14,500 feet MSL up to but not including 18,000 feet MSL. In other words, Class E airspace bumps up against the bottom of Class A airspace.[59] But I used

---

[59] Class E airspace is also *above* Class A airspace, which ends at 60,000 feet. But unless you're in a helium-filled balloon or a spy plane of some kind, you needn't bother yourself with this fact!

the word "standard" when referring to the 14,500 feet MSL lower boundary of Class E airspace because like a lot of standards in aviation, it's a rule with a ton of exceptions.

On a sectional, Class E airspace, as the definition we referenced earlier tells us, is basically everywhere that Class A, B, C, and D isn't. But because the floor of Class E varies, the map-drawers use colors, dashed lines, and "fuzzy" lines to indicate the lower boundary of Class E airspace in various areas. The most prominent of these symbols is the magenta line that fades on one side—the "fuzzy" part. This is the most prevalent of all markings on a sectional chart. It's *everywhere* (helping Class E airspace live up to its "everywhere" nickname). Zoom out on a digital sectional and you'll see Class E markings all over the map like a rash on the skin.

To aid in our discussion of Class E airspace, I'll be bouncing around the country on a sectional map so you may want to pull up an electronic version of one to follow along. Let's start at KCRS—C. David Campbell Field, also shown as Corsicana Municipal. Some of my extended family live on a ranch near here where I spent many summers as a kid, so it has a special place in my heart. It also helps us get a good start on understanding Class E airspace. You'll notice that there is one of those magenta solid-to-fuzzy circles around the airport. On the fuzzy side of that line, the Class E airspace goes all the way down to 700 feet AGL. Now that you know that pilots on IFR flight plans are controlled when they are in Class E airspace, you probably can guess why the "floor" of the Class E around an airport might be lower than the "standard" 14,500 feet MSL. It's because pilots on instrument approaches and departures into and out of uncontrolled airports still need the benefits of ATC control for traffic separation until they get close to the runway.

A quick aside: note that while the "standard" bottom of Class E airspace is 14,500 feet MSL, the fuzzy magenta outline around KCRS denotes the floor at 700 AGL—i.e., *above ground level.* The Class E floors closest to the surface of the Earth (we'll cover the other altitudes in a minute) are relative to the ground, not sea level. This is another thing that can be a bit confusing at first blush. But when you consider it more deeply, it makes sense. Because all airports vary in their altitude above sea level, defining the floor of Class E airspace very near to the ground would not be very useful if done in MSL. If MSL were used at KCRS for example, the bottom of the Class E would be 1,149 feet since the airport altitude is 449 feet MSL (449+700=1,149). If you went 24 miles to the south of KCRS to KLXY, an airport with an altitude of 545 feet MSL, the bottom of the Class E would have to be designated as 1,245 feet MSL. It's a lot easier to use AGL rather than try to think through the MSL altitudes at every airport in the country. The exception to this AGL rule is the "standard" Class E bottom: 14,500 feet. This is measured in MSL.

Back to KCRS....

Outside of the magenta circle around KCRS—on the "solid" side of the line—the bottom of the Class E is 1,200 feet. If you look again at KCRS, you'll notice the Cedar Creek VOR to the northeast.[60] This VOR sits beneath Class E airspace with a floor of 1,200 feet AGL. Again, it makes sense that the Class E floor would be higher in an area like this because there is no airport in the immediate vicinity. But it does beg the question: what's

---

[60] VOR stands for "VHF Omnidirectional Range." A VOR station is a navigation aid that you'll learn to use during your training. A VOR intersection is a point in space where two radials from two different VORs intersect. Radials define specific headings to or from a VOR.

special about 1,200 feet AGL? It's special because victor airways begin at 1,200 feet. Victor airways can be thought of as highways in the sky that pilots on IFR flight plans at lower altitudes use. When I say "lower altitudes," I mean from 1,200 feet AGL up to but not including 18,000 feet MSL—the bottom of our Class A airspace. Victor airways run between VOR stations or VOR stations and VOR intersections. If you look at the Cedar Creek VOR just northeast of KCRS, you'll see light blue lines radiating from it. These are Victor airways (designated by the letter *V* followed by a number).

The next possibility to consider for the bottom of a Class E airspace is the surface. Class E will extend all the way down to the surface when an airport has a precision instrument approach. On a precision approach, an airplane has both vertical and horizontal guidance to the runway. This means that a pilot on an IFR flight plan can get a lot closer to the ground before they need to see the runway (and in some cases don't need to see it at all!). Said another way, a plane on a precision instrument approach can be in the clouds or very low visibility at *very* low altitudes. Knowing this helps us understand why the Class E might extend to the surface—to allow ATC to provide separation for IFR traffic in circumstances where planes can be flying very close to the ground near an airport.

In cases where Class E goes all the way to the surface, you will see a dashed magenta line. To explore an example of this, zip on over to KACK on your electronic sectional and you'll find yourself just south of Cape Cod, Massachusetts, on Nantucket Island. You'll probably first notice that there is a shaded magenta line around the airport defining Class E down to 700 feet AGL. You'll also notice that unlike KCRS, this airspace is not just a circle around the airport. There is an extension off the circle to the northeast. Want to guess why this is here? You'll notice that it happens to line up with the Runway 24. Because there are instrument approaches to Runway 24, this helps protect IFR traffic descending to KACK while on those approaches. Inside of the shaded magenta border there is something else—the dashed magenta line I referenced earlier. This defines the area where Class E goes all the way to the surface because there isn't just an instrument approach to Runway 24, but a *precision* instrument approach to Runway 24. If you think of the 700 feet floor and the surface floor together, you've essentially created a mini upside-down wedding cake with two layers, kind of like a Class C airspace. The difference is that VFR traffic can fly in and out of the Class E without any permission or without maintaining contact with ATC (assuming weather minimums are met—more on this in a bit). That's something you can't do when it comes to Class B, C, and D airspace.

We've established that Class E airspace tops out at just under 18,000 feet MSL and has three very low options for bottoms: the surface, 700 feet AGL, and 1,200 feet AGL. Where, then, is the 14,500-foot MSL floor used? When you take a broad view of a sectional, it appears Class E everywhere is either at 700 feet AGL (inside our fuzzy magenta boundary) or 1,200 feet AGL (outside our fuzzy magenta boundary). And it appears that way because *so* much of Class E airspace does have one of these two altitudes as a floor. Continuing with our effort to understand why airspace is divided up the way it is, can you think of what would have to be true for the Class E floor to be up at 14,500 feet MSL?

Since the surface, 700 feet AGL, and 1,200 feet AGL floors are related to providing control to IFR pilots on a Victor airway or during an instrument departure or procedure to an airport, it makes sense that the 14,500 feet MSL floor must be in a place where there is no need to control pilots on IFR flight plans closer to the ground. In other words, a place where there are no airports with instrument procedures and no Victor airways crossing overhead. Where might that be? If you guessed areas of expansive nothingness, you'd be right. And in the U.S., that means you'll be looking at areas in the western half of the country.

Let's do that now and head out to Arizona. Find KJTC, Springerville Municipal Airport. You'll note that this airport sits inside of a magenta ring just like KCRS, so it has a Class E floor of 700 feet AGL. But we're not interested in this airport, we're interested in what's to the east of it. If you look in that direction. you'll notice a (sort of) rectangular shape outlined by a dark bluish border. That border is fuzzy on the outside of the shape and solid on the inside. The area on the solid side—i.e., inside the shape—is Class E airspace that has a floor of 14,500 feet MSL; under that, you're in Glass G airspace. On the fuzzy side of the shape, you're back in Class E with a floor of 1,200 feet AGL. There are no airports under this blue-border-encapsulated airspace. There are no Victor airways through it. As the country becomes more populated, Class E with the 14,500 feet MSL floor has become rarer. And for that reason, most of you will never be flying under it while cruising along at 3,500 feet (or any other altitude up to 14,500 feet MSL).

Class G
Fills in the gaps

Class G airspace is defined by the FAA as "that portion of the airspace that has not been designated as Class A, Class B, Class C, Class D, or Class E airspace." Yikes ... another definition that defines something by saying what it's *not.* If you're looking for some help in understanding what it *is*, then remembering that it's the only *uncontrolled* airspace may be helpful. This only means that neither VFR nor IFR pilots need a clearance to fly in Class G airspace. Finally, when thinking about where Class G is beyond saying "everywhere Class A, B, C, D, and E airspace isn't," note that Class G airspace *always* lies under Class E airspace. It goes from the surface to the floor of the overlying Class E airspace which, as we know by now, means it goes from the surface to either 700 feet AGL, 1,200 feet AGL, or 14,500 feet MSL (obviously, if Class E goes to the surface, then there can be no Class G under it).

Hopefully, understanding the logic in how the sky above you is partitioned by regulators helps you remember the nuances of each of the categories of airspace. I know it does for me. But as a VFR private pilot, it's not enough to know the dimensions of the airspaces, you also need to know the minimum cloud clearances and visibility required to fly inside of them.

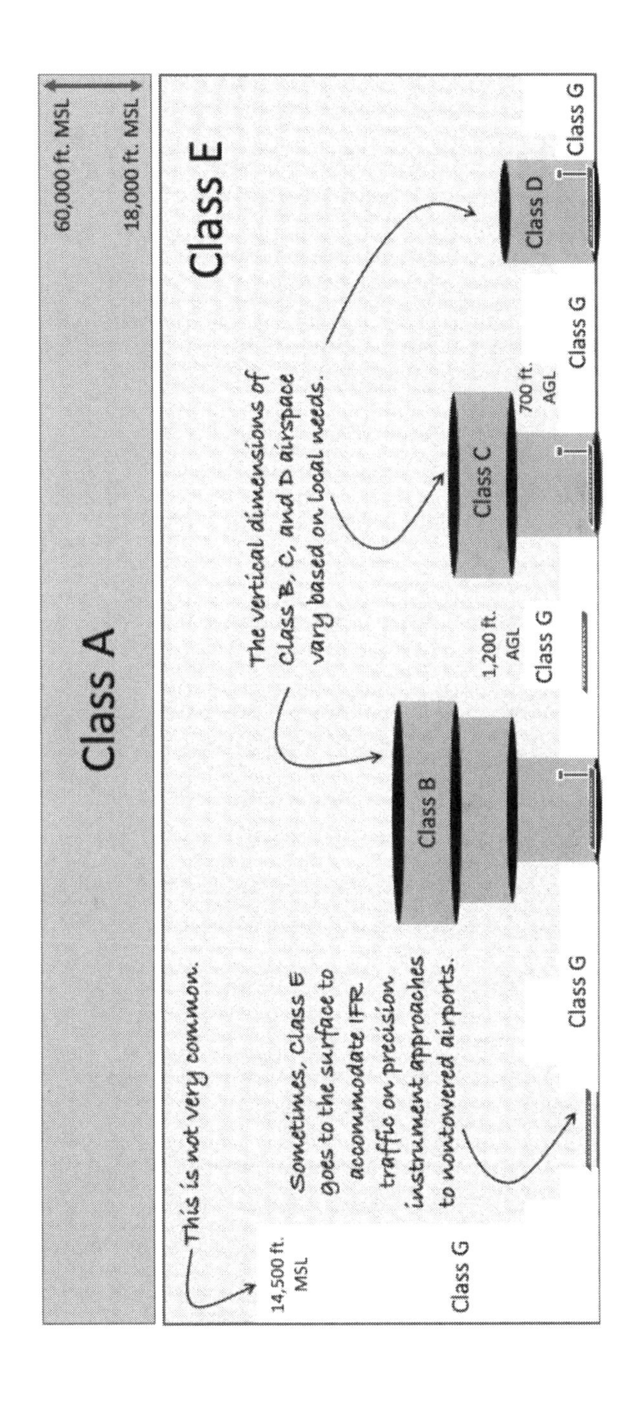
60,000 ft. MSL
18,000 ft. MSL
Class A
Class E
14,500 ft. MSL
This is not very common.
Sometimes, Class E goes to the surface to accommodate IFR traffic on precision instrument approaches to nontowered airports.
The vertical dimensions of Class B, C, and D airspace vary based on local needs.
Class B
Class C
Class D
1,200 ft. AGL
700 ft. AGL
Class G
Class G
Class G
Class G
Class G

Cloud clearance and visibility requirements

Unfortunately, when it comes to cloud clearance and visibility requirements, there's not a lot of consistency in how the regulators have defined those requirements. While that's frustrating, like the design of airspace itself, there's some method to the madness.

Before getting into those methods, it's good to know that while the requirements vary quite a bit, they are all a combination of the following restrictions:

Visibility—5 sm, 3 sm, 1 sm
Distance above a cloud—clear of cloud, 500 feet, 1,000 feet
Distance below a cloud—clear of cloud, 500 feet, 1,000 feet
Distance laterally from cloud—clear of cloud, 2,000 feet, 1 sm

If we boil those numbers down to just the first digits, then this list looks like this:

Visibility—5, 3, 1
Distance above a cloud—Clear, 5, 1
Distance below a cloud—Clear, 5, 1
Distance laterally from a cloud—Clear, 2, 1

With this shorthand in mind, you can now imagine a visual aid that would help you remember, for example, the separation required from clouds. In fact, other than staying clear of clouds (a requirement in some airspaces at certain times), the only other two options for cloud clearances look like this:

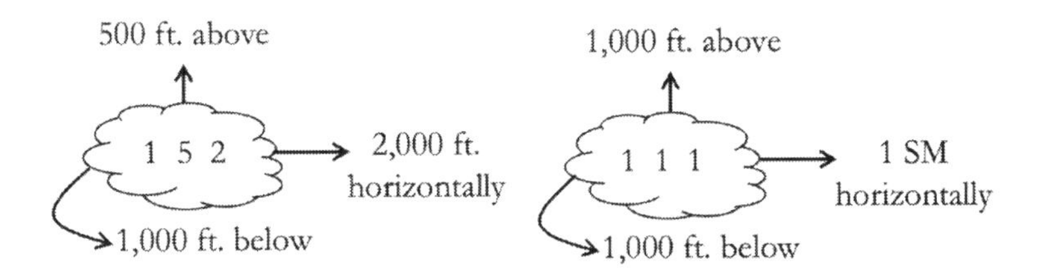

And there are only three options for visibility:

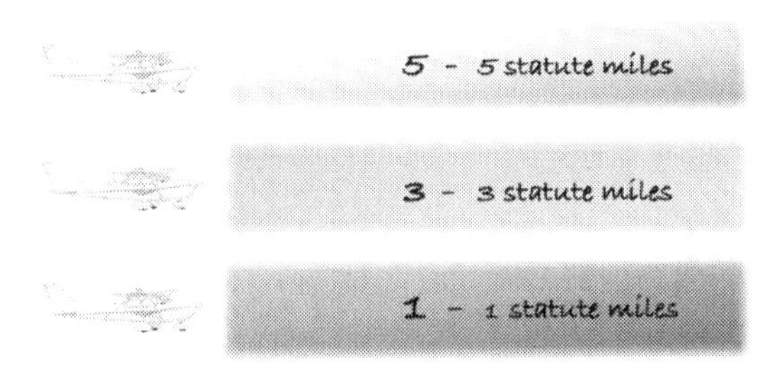

Now that we have our possibilities for visibility and cloud separation defined, we can draw a graphic that puts it all together with airspaces to help us remember what's required where and when. There are many variations on my own graphic below but fundamentally they all use this triangles-within-triangle structure:

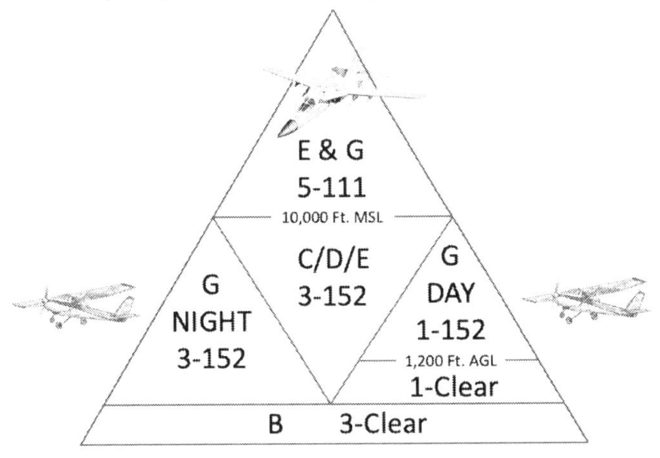

Since we laid the groundwork for the options, you probably have already figured out how to read this. But a little explanation might help. Look at the triangle in the center. This area tells you that when in class C, D, and E airspace below 10,000 feet MSL, you must have 3 sm of visibility and remain 1,000 feet above, 500 feet below, and 2,000 feet to the side of a cloud (152). I've added a sketch of a Cessna 152 on either side of the lower part of the large triangle to help you remember the 152 requirement that is used frequently. The top triangle says that when in E and G airspace above 10,000 feet MSL, you must have 5 sm of visibility and remain 1,000 feet above, 1,000 feet below, and 1 sm to the side of a cloud. I've put a sketch of the F-111 fighter/bomber, a plane my father flew in (and ejected from—more on that later) as a reminder of the 111 requirements structure in this airspace.

While I found a triangle like this very helpful as it was easier for me to memorize than a table with the same values, I'd suggest that taking the time to understand the logic of the requirements will aid memory significantly. At the core of the requirements is one principle: VFR aircraft must always be able to "see and avoid" other traffic. Recall that in all but one of our airspaces (Glass G), IFR traffic will be controlled by ATC. Because you, as a VFR pilot, won't be, you need to be sure to have time to see other aircraft if you are to avoid them. The visibility and cloud clearance requirements are all about giving you that time.

You'll notice that in Class G airspace below 1,200 feet AGL (where most Class G airspace is in the U.S.) you have the least restrictive requirements: 1 sm visibility and clear of clouds. In this airspace, the assumption is that this is enough room to give you time to avoid another aircraft. Aircraft at this altitude are generally flying more slowly. Somebody on an IFR flight plan transitioning out of controlled airspace into uncontrolled Class

G who is below this 1,200 feet AGL threshold will likely be landing at an uncontrolled airport and therefore will be slower too and will also be making radio calls to deconflict with other traffic.[61]

Contrast this with requirements above 1,200 feet AGL and below 10,000 feet MSL in Class G, and all Class E airspace below 10,000 feet MSL. In these airspaces, VFR pilots must maintain the 152 cloud clearance requirements. They must also have at least 3 sm of visibility (in Class E all the time and Class G at night; it's 1 sm during the day in Class G). You can probably guess why the requirement is more stringent at these higher altitudes. Aircraft are flying faster and, in the case of Class E airspace, IFR traffic will be flying in and out of clouds. By staying further away from those clouds and having more visibility, you'll have more time (as will the other pilot) to avoid a collision if necessary. Above 10,000 feet MSL in both Class E and Class G, you must stay even farther from clouds. You'll be among more IFR traffic here and the speeds will be even higher.

When it comes to Class C and D airspace (remember, as a VFR pilot, you can't be in Class A), the requirements are the same as they are for Class E below 10,000 feet MSL for all the reasons we've already discussed. Class B is different though. The requirements are *less* stringent than they are in Class C and D airspace, at least when it comes to cloud separation. In Class B, you just need to stay clear of clouds. When you're inside of Class B airspace, even as a VFR pilot, you are "positively controlled" by ATC. They are ensuring that you are separated from other

---

[61] While an IFR pilot might be making radio calls, depending on where they are landing, other traffic may not be since radios are not required in many airspaces in the United States.

traffic.[62] This means that you don't have to worry about somebody on an IFR flight plan popping out of a cloud right next to you; ATC should know where everybody is and ensure that this can't happen. That said, I must emphasize again that you should always be taking the time and effort to see and avoid.

Speaking of see and avoid, I wondered if the requirements around visibility and cloud separation had any science to them. While researching this question, I came across an interesting FAA study from 1971 titled "Analysis of VFR Cloud Clearance and Visibility Standards." Despite its age, it's still a relevant read and addresses a variation on my question of the science of the restrictions represented in our triangle graphic. The authors ask: Are the visibility and cloud clearance requirements sufficient to ensure pilots have enough time to react in an effort to see and avoid?

Before delving more deeply into that question, the authors, George Rowland and Carl Reichwein, discuss what we have been touching on in this section and then point out that see and avoid seems to work:

> *These [visibility and cloud clearance] requirements have been based upon the general appreciation of the fact that a pilot needs some minimum amount of time to detect the presence of, and maneuver away from, another aircraft which has just loomed out of the haze, come around from behind a cloud, or come up or down at him out of a cloud layer. All things considered, the concept works surprisingly well.*

---

[62] Recall that if you get flight following and are flying outside of Class B airspace, you aren't positively controlled. ATC does not have an obligation to separate you from other traffic. They will just advise as workload permits.

Their last point is a good one. While there have been plenty of mid-air collisions in recent years, I can't think of a single one due to an IFR pilot, for example, popping out of a cloud and running into a VFR pilot who was flying too close to that cloud. In fact, every incident I can recall has been in good visibility with few clouds in the immediate vicinity of an airport. This is, in fact, the most likely time for a midair collision to occur based on available data.

The authors do point out though that as aircraft performance has increased, the limitations have become less effective. After all, suppose an IFR jet doing 200 knots penetrates a cloud that you're on the other side of. In this scenario, even if you are 1,000 feet away from the side of that cloud, the jet will get to you in just under three seconds—and that assumes no portion of your direction of travel is toward the IFR jet. If you were flying even a little in that plane's direction, the closure rate would be even faster. That's very little time for either pilot to react.

In light of this, why not make the separation requirements more stringent? As the authors of the report point out, doing so will make it harder for VFR pilots to get in the air. Imagine a sky with clouds separated by 2,000 feet on all sides in Class E airspace. Under current rules, a VFR pilot could fly between those clouds and be exactly separated 1,000 feet from each. (In practice, this would be virtually impossible to do, so if you're a stickler for the details, bear with me in this thought experiment.) Now imagine the regulation was changed to 1,500 feet of clearance between clouds to allow more time for pilots to see and avoid. If this rule change was made, then a VFR pilot could no longer fly between clouds separated by 2,000 feet. Those clouds would have to be at least 3,000 feet apart from each other.

"But the pilot could stay under the clouds," you might argue. True—but if you are increasing the required lateral separation from clouds, then you are probably increasing the below-clouds separation requirements. This has the effect of "squeezing" the pilot closer to the ground. Not ideal if you are trying to make the sky VFR friendly.

I couldn't find when the original cloud clearance and visibility requirements were codified in the regulations but suffice it to say that based on the Rowland and Reichwein report, it must have been well before 1971. And while aircraft speed—in both commercial and general aviation—has increased in the decades since, new technologies have made it easier to ensure aircraft stay separated.

ADS-B has done a lot in this regard. "Automatic Dependent Surveillance—Broadcast" is (per the FAA) an "advanced surveillance technology that combines an aircraft's positioning source, aircraft avionics, and ground infrastructure to create an accurate surveillance interface between aircraft and ATC." In short, it makes it much easier not only for ATC to get a complete picture of a given airspace but also to allow individual pilots to as well.

Early in my flight training, I purchased a portable ADS-B receiver that connects to my iPad and shows me not only my own position on a moving map along with any flight plan I've created, but also shows me other traffic (note that all the aircraft I trained in also had ADS-B transmitters sending a signal "out" to report their position and altitude). Admittedly, ADS-B isn't on all aircraft. But in the airspace I often fly in—within 30 miles of the Boston Class B—it's required.[63] ADS-B helped me avoid some close calls with other aircraft at least three times over about a two-year period. The required use of ADS-B inside of crowded airspaces has undoubtedly helped make the sky safer and has therefore allowed the visibility and cloud clearance requirements to remain as liberal as they are.

While we're on the subject of ADS-B and see and avoid, a short rant. I would love to see the FAA require "ADS-B out" on *all* aircraft, regardless of the airspace they operate in. The dirty little secret is that "see and avoid" is exceptionally hard, bordering on impossible. Old-timers who bad-mouth the use of technology to identify nearby aircraft (and I've run into several) are simply wrong to say that using your eyes alone is the best way to go. How do I know? *Even when I know where traffic is based on ADS-B information, sighting it is virtually impossible.* And I know this isn't some shortcoming of mine; I've flown with plenty of passengers and other pilots, and they all have the same difficulty. I hear pilots regularly say they are unable to see traffic ATC is pointing out to them. Part of the argument traditionalists make for keeping your eyes outside the cockpit all the time is that there may be an aircraft near you that doesn't have an ADS-B transmitter. Well of course … so why don't we require all aircraft to have transmitters so that we remove some of this risk? I understand the desire to keep flying inexpensive (retrofitting an aircraft with ADS-B will cost several thousand dollars in most

---

[63] ADS-B is also required inside of Class A, B, and C airspace, in Class E above 10,000 feet (excluding airspace at or below 2,500 feet AGL), above the ceiling and within the lateral boundaries of Class B or Class C airspace up to 10,000 feet, and in Class E airspace over the Gulf of Mexico at and above 3,000 feet MSL within 12 nm of the U.S. coast.

cases and not all aircraft even have electrical systems), but should that be our primary concern, particularly when the hobby is an expensive one to begin with?

If you think I'm being overly dramatic, several weeks before I wrote this paragraph, a midair collision in Florida killed four people, two of whom were in a Piper aircraft and two of whom were in a J-3 Cub floatplane. The aircraft were flying near Winter Haven Regional Airport (KGIF) —the Piper was doing practice approaches—when they collided nearly head-on. The Cub had no radios or an ADS-B transmitter (and it wasn't required to by FAA regulations). Even if the Piper student pilot and instructor had in-cockpit ADS-B traffic displayed (and at the time of this writing, that was unknown), they would never have known the Cub was near them. What's sadly ironic is that KGIF sits just 2.5 miles to the west of the Orlando and 12 miles east of the Tampa "Mode C veils"—the boundaries that circle these Class B airports and within which aircraft are required to have ADS-B out. Can there be any doubt that this corridor between these two airports in sunny Florida is one of the busiest general aviation slices of the sky? In fact, the arrangement of airspace here has the effect of funneling all aircraft without ADS-B traversing the length of the state into this area, since those Mode C Veil boundaries for Tampa and Orlando stretch all the way to the Gulf of Mexico in the west and nearly all the way to the Atlantic in the east. It just seems insane that we have technology to greatly improve safety at a reasonable cost and yet we play with fire. Assume the cost of installing ADS-B in the Cub was $20,000. Do you think the families of those killed would cough up that money to get their loved ones back? Something tells me the answer is "yes."

Special use airspace

In addition to the airspaces discussed thus far, there's one other category of airspace to be aware of: special use airspace (SUA). These airspaces either restrict GA pilots from flying within their boundaries or warn them of dangers that may exist within the boundaries—like the possibility of getting shot down. As with other airspaces though, the regulators (at least in the United States) do all they can to ensure that GA pilots can use most of the sky even if it needs to be shared with national defense activities or security operations. In fact, most of SUA is accessible to GA pilots. But entering SUAs when you shouldn't can have significant, adverse consequences.

MOAs – don't get shot down

The most common form of special use airspace is MOAs—military operations areas. If you go to your digital sectional and find St. Mary's Municipal Airport in Pennsylvania (KOYM), you'll find yourself under an MOA: the Duke MOA. The boundaries of the MOA are identified by a magenta line with "spikes" facing toward the inside of the bounded area.

I use ForeFlight as my EFB. The application allows you to "long tap" within the MOA and a list of airports, airspaces, waypoints, etc. appear grouped into "Locations" (e.g., the GPS coordinates of the exact location you tapped), "Airspace" (e.g., the MOA we are discussing), and "Nearby" (e.g., nearby airports and waypoints). Tapping once again on the "Details" button to the right of the MOA name gives you the particulars regarding hours of use, the frequency to contact to ask about whether it is active (or "hot"), and the elevations bounding its top and bottom. If you're old-school, you can find this information along the edge of a sectional in a table detailing all SUAs that appear on that sectional. The Duke MOA, for example, is shown to have hours of operation that are "non-continuous." Active hours are typically between 1000-1500 universal coordinated time (UTC) daily, but it may be in use outside of these times.[64] When that is the case, the FAA will notify pilots through NOTAMs (Notices To Air Missions). The lower limit of the MOA is 8,000 feet MSL and upper limit is 17,999 feet MSL. You probably notice that upper limit as the bottom of Class A airspace. Best to keep that military activity away from all those airliners crisscrossing the continent. Finally, if you want to ask about the status of the MOA—whether it's "hot" or not—you can call up Cleveland Center on 124.32 or 125.87.

Like much of the rest of the airspace system, MOAs are built with IFR traffic in mind (are you starting to get a VFR inferiority complex?). The FAA notes that MOAs are "established for the purpose of separating certain military training activities from IFR traffic." IFR traffic is allowed to be cleared through the MOA if separation between aircraft can be provided by ATC. What's interesting is that while MOAs are made with IFR traffic in mind, they are open to VFR traffic. Even when they are "hot." That's right—you can fly through an MOA even if military operations are happening. But keep in mind that you do this at your own risk. The FAA points out that activities in MOAs include things like air combat tactics training, air intercepts, aerobatics, formation training and low-level flying. And, unlike the rest of us, those aircraft are authorized to fly at speeds exceeding 250

---

[64] UTC is a time fixed across all time zones. It provides a common time for everybody in the aviation community to use. No matter what the local time is, UTC is fixed globally. In military applications, UTC is often called "Zulu" time.

knots below 10,000 feet MSL. So, see and avoid might be a tad more difficult. Don't be shy about using the frequencies listed to ask if an MOA is hot or not. While I've never had to do this, I've heard other pilots on frequency while I've been flying nearby do so and controllers have always been helpful and informative.

Restricted airspace – the aliens may get you

Restricted airspace is not as accommodating to VFR pilots as MOAs. Let's pick a fun, and very famous example of restricted airspace to advance our learning: Area 51. Many of you may know this as the mysterious military facility north of Las Vegas, Nevada near a dry lakebed known as "Groom Lake." The location is believed to be the place where U.S. military and intelligence organizations test new technology that they wish to keep from the prying eyes of curious civilians like you and me and, more importantly, adversaries around the world. It's also the place where many (I don't count myself among those many, by the way) think the U.S. government is hiding alien remains—or maybe even live aliens.

When looking for the location of Area 51 on a sectional, I learned that it has an identifier that an EFB like ForeFlight recognizes: KXTA. This identifier is associated with an airport named "Homey." But on ForeFlight's electronic sectional, no airport is shown when the application takes you to the location the identifier specifies. There's just a green dot placed on the map south of Groom Lake. And indeed, I'm unable to find any official record of KXTA. A search of an FAA airport database returns a terse "no results found." Interestingly, ForeFlight does provide a diagram of the airport that shows a single hard-surface runway,

14/23, and two other runways on the surface of dry Groom Lake. There are no FBOs listed and no tower frequencies.

Using Google Maps, I was able to confirm that the location of the green dot is indeed the location of the airport at the center of Area 51. You can clearly see Runway 14/23, paralleling a closed runway, *Xs* painted along its full length. The open runway is around 12,000 feet long. That's not all that long given the altitude of the airport, around 4,500 feet, and the fact that it regularly sees high temperatures in the summer months that make the density altitude quite high. But I guess when you are using dilithium crystals and flux capacitors to get off the ground, you don't need a long runway.

The suspicion regarding why KXTA shows up in private aviation databases and not official government ones (well, not *publicly available* ones) is that the information regarding the airport was accidently released in 2007. Some surmise—I think correctly—that the KXTA designation is a tip of the hat to the word "extraterrestrial." If so, well played CIA. Less clear is why the name of the airport is "Homey." I can't find *anything* regarding where that name might have come from. I searched to see if there were any military aviators with that name who had perished while flying. As I mentioned earlier in the book, Air Force bases are almost always named after aviators who died in the line of duty. But nothing came up. Of course, if an aviator with that name died while testing something classified at Area 51, the cause of death may not be public. Maybe it's simply ironic—from a satellite image, nothing looks "homey" about the location. Desolate comes more readily to mind. This would be consistent with an early name for the location, Paradise Ranch, which was supposedly meant to encourage workers to move there. This was clearly in the days before the internet allowed the truth to be revealed.

One thing is for sure, the name "Homey" isn't meant to be inviting to general aviation pilots. We went down this rabbit hole while discussing restricted areas and KXTA lies under one: R-4808 N. You can identify its boundary by a blue border comprised of a solid line with "spikes" to the inside of the bounded area (the markings are identical to those of an MOA, just a different color). The lower limit of the R-4808 N airspace is "surface," and the upper limit is "unlimited." The space is active "continuously (incl holidays)" and the hours are "continuous." None of these restrictions leave very much room for you to squeak through the airspace in your Piper Cub.

Like MOAs, restricted airspace is available to general aviation traffic. But there are some hurdles, particularly if you are flying VFR. A VFR pilot must get permission in advance to fly through restricted airspace. And if that space is hot during your time of planned transit, your likelihood of approval is about as high as the likelihood that aliens are at Area 51. Even when the area is "cold" you may not get permission. The key is to know that you must have permission to enter this space, unlike an MOA. Aircraft flying IFR flight plans have it a bit easier. ATC will clear traffic through these airspaces if the owner of the airspace has released it—i.e., made it available to other traffic.

Prohibited – the name means what it says

Let's now head out to KFDK, Frederick Municipal Airport, a Class D airport northwest of Washington, D.C. It sits right between D.C. and Camp David, the presidential retreat. You can see easily where Camp David is; just north and slightly west of

KFDK there is a circle designating both restricted airspace (R-4009) and prohibited airspace (P-40). This is because one sits on top of the other. The prohibited airspace goes from the surface to 4,999 feet MSL and is about six miles in diameter. The restricted airspace goes from 5,000 feet MSL (the top of the prohibited airspace) all the way up to 12,500 feet MSL and is about 20 miles in diameter. With the addition of the restricted airspace on top of the prohibited, the FAA created what looks like a two-layer Class C airspace over Camp David. The notes for the restricted area say that it is activated continuously (including holidays) and that the hours are continuous as well—which, like Area 51, is redundant but clear.

All prohibited airspace has a "floor" that is the surface of the Earth. It's written into the definition: *a prohibited airspace begins at the surface and has defined dimensions in which the flight of unauthorized aircraft is prohibited.* The reason for this is that prohibited airspace is typically about protecting something on the ground (like Camp David and the President of the United States). Restricted airspace is usually about protecting aircraft in the air—e.g., keeping you from being hit by a surface-to-air missile. The Camp David example is one where this is not strictly true. The restricted airspace is more about extending the protection around the prohibited area to allow authorities more time to react to a potential threat if a VIP is at Camp David. Of course, if you violate the airspace without permission, you may get shot down by a missile in any case!

While "prohibited" sounds more foreboding than "restricted," the rules around both are similar. For example, believe it or not, you *can* fly through prohibited airspace *if you get permission.* The legal language in 14 CFR § 91.133 says the following: *No person may operate an aircraft within a restricted area …*

contrary to the restrictions imposed, or within a prohibited area, unless that person has the permission of the using or controlling agency, as appropriate. Of course, it's the getting permission part of the regulation that makes flying through a prohibited area, for all intents and purposes, impossible as a GA VFR pilot.

There are a couple of other special use airspaces (warning areas for example) but MOAs, restricted, and prohibited airspaces are the ones you're most likely to encounter in your jaunts to get a $100 hamburger. If your route of travel does indeed take you through or near one of these areas, be sure to re-plan accordingly or get the permission you need. If you don't, you could end up in prison…or worse.

Chapter 17
The Solo Cross-Country Flight

With knowledge of how to plan a cross-country flight and how to navigate airspace, I did my first cross-country flight with Connor in August, six months after my discovery flight. I flew one other such flight with Jeff in May, so this wasn't my first time to venture out of the pattern or beyond our practice area. But by August I was feeling a lot more confident in my ability to conduct a long flight to another airport by myself. I had 4.2 hours of solo time and about another 24 hours of additional "dual received" by the time my first cross-country with Connor took place.

Two days after that daytime flight where we practiced pilotage, dead reckoning, and a diversion, we did our first nighttime cross-country, on August 26. I had read that one way to effectively fly and conduct a preflight while in the dark was to wear a headlamp, so I bought one with both a white light and red light, the latter color helping to preserve nighttime vision. Connor laughed at me when he saw me wearing it. I suspect he thought I looked like a dork. And I probably did. But I didn't think fashion was a priority.

But feeling well when flying is. And, while it took me a while to figure it out, the headlamp was giving me a raging headache.

If I have one headache a year, I'd be surprised. Even when I have had the flu or some other illness, rarely does it manifest in the form of a headache. But during our flight to Sanford, Maine, that evening, I was feeling horrible; the more my head hurt, the more nauseous I became. In retrospect, I'm surprised I didn't suspect carbon monoxide poisoning. But it wasn't anything to do with the airplane. The strap on the headlamp was too tight. It was strangling my noggin. I only thought to remove it about two-thirds of the way into our adventure and got instant relief. If you decide to use a headlamp for night operations, be sure to wear it around the house for an extended period to check comfort and adjust the strap accordingly. You'll look like a misplaced coal miner for a while. But better to look out of place while doing chores around the house than be as distracted as I was in the air because you feel like your head is going to explode.

The night of our flight was typical for August in New England. It had been hot that day: 96° F. And the dew point peaked at 74° F, making the air very, very sticky. What was common for a month in New England is common for half the year in much of the rest of the country when heat and humidity frequently combine to make the air a milky soup. Visibility was officially greater than 10 miles when we took off with much of the same expected along our route, and we had a cloudless sky. The fact that the sky was cloudless, though, didn't mean we could see stars. And I doubt we could even see the 10 miles the weather report for Hanscom said we could.

The route we flew north to Maine on that nighttime flight took us along the coast of the Eastern Seaboard. At one point, Connor said, "My controls," and I responded with, "Your controls." He followed once again with "My controls." This back-and-forth is something I'm sure you've done with your instructor. It's meant to ensure there is no confusion around who is flying the airplane. Be sure you pay attention to this and relinquish the yoke, throttle, and rudder pedals if your instructor asks. While writing this book, there was an accident that killed a CFI and his student when the student apparently locked up his hands and arms while the stick (this airplane did not have a yoke) was held in a full back position, putting the aircraft into a fatal stall/spin. Because somebody (likely the student) was pressing the mic button on the top of the stick during the incident, audio transmissions were captured of the CFI yelling at the student to let go of the controls. But it was to no avail. If your CFI is asking for the controls, there's a reason.

Connor's reason for asking me for the controls was less urgent than the CFI's in that accident scenario. But it was related to demonstrating something to me in the interest of safety.

We were on a 220° heading returning from KSFM (Sanford, Maine) back to Hanscom, having completed a landing a couple of minutes before. Connor turned the aircraft to a 90° heading, due east. We flew for a short while until the shoreline disappeared from our view and the Atlantic became the predominant feature in front of us. Well, predominant as far as I knew. Because I couldn't see it. I couldn't see anything. And that was the point.

Connor wanted to show me how even on a legal VFR flight, one that met visibility and cloud cover requirements handily, you could find yourself in what were essentially instrument conditions. The haze and lack of a moon made the dark of the ocean melt into the dark of the sky. It was as if we were flying through space, not off the coast of New England. Connor was an instrument-rated pilot, so I felt safe with him at the controls in this scenario. But even so, I locked my gaze on the attitude indicator to be sure we were staying straight and level.

The most famous crash to illustrate the danger Connor was demonstrating occurred almost exactly 100 miles to the south of our position that night. John F. Kennedy Jr., the son of the former president, crashed off the coast of Martha's Vineyard while piloting a Piper Saratoga killing himself, his wife, and his sister-in-law. Like many accidents, this one had the "Swiss cheese" model at play. That model says that several "holes" in a sequence of events must line up for an accident to happen; it's never just one thing. So it was on that fateful day in July 1999.

I can't get inside of JFK Jr.'s head, but in reading much about his flying history, it's hard not to think that he believed his celebrity status was some sort of inoculation against bad things happening to him. About three weeks before he fired up the engine on his airplane for the last time, his left ankle was cut from a cast (he broke it after a paragliding or hang gliding accident; there are conflicting reports on this). The injury was severe enough to require surgery, and he had only been able to stop using a motion-limiting boot to walk just days before the accident. On the day of the accident, he was still using a cane to assist his mobility. Did his survival and recovery from the paragliding incident reinforce his sense that he was invincible? Maybe. His own wife had confided in somebody that she feared flying with JFK Jr. because he didn't seem to take it seriously enough. If that's not a sign of invincibility, I'm not sure what is.

He was apparently in a hurry to get to where he was going. He had a wedding to attend the next day. But his departure was delayed by the late arrival of one of his passengers. It was a perfect setup for "get-there-itis," something that has been a contributing cause of death for more pilots than I'd care to count. He was flying an aircraft that he wasn't too familiar with and which he often flew with an instructor. But he'd be solo on this night flight. He would be solo in another regard; he never requested flight following. Doing this wouldn't necessarily have saved him—after all, a controller can't fly the airplane—but he might have been advised by controllers of pilot reports (PIREPs) detailing the status of weather along his route. Even if he didn't get the reports directly, just being tuned into a frequency for the area would have allowed him to hear any commentary from others about the conditions they were encountering.

That last point about the conditions brings us back to why Connor had turned our plane to the east and flown to a location where no ground references were available. Ultimately, this was the final hole in the Swiss cheese for JFK Jr. He flew into a very thick layer of haze that day, one that would have been the functional equivalent of flying into a cloud. The haze was that bad. This point is made well in a web article by pilot John Emmerling.

Titled "How 4 Avoidable Mistakes Doomed the Final Flight of JFK Jr.," the article tells the story of how Emmerling piloted a flight along the same route on the same day and close to the same time as JFK Jr.'s final flight. Emmerling departed KHPN (White Plains, New York) for KCQX (Chatham, Massachusetts) a couple of hours before JFK Jr. left Essex County Airport (KCDW). He begins by saying that the flight conditions that day were "hot and hazy with very low visibility." An accomplished instrument-rated pilot, Emmerling had filed an IFR flight plan, so the "soup" wasn't a problem for him. Still, he was surprised by how thick and persistent it was, much more so than usual. He attempted to climb his aircraft (a high-performance Piper Malibu) above the limiting haze, but even after "reaching 9,000 feet, the visibility outside the cockpit was still—*still!* —zero, zilch." It was bad.

According to the NTSB report on the accident, JFK Jr. (or somebody using his login code) did request some weather information from a computer terminal at the airport. And in his defense, reports around the time of his departure didn't seem too daunting. His final destination, Hyannis, was reporting few clouds at 7,000 feet, visibility six miles in haze, and winds 230 at 13 knots (note that KHYA has a Runway 24, so these winds would have been almost right down the runway). Martha's Vineyard, his stopover, was showing even better weather: clear skies, visibility six miles with haze, and winds 210 degrees at 11 knots (KMVY also has a Runway 24).

But the time with Connor that night showed me how a good weather report can be very misleading under certain conditions. It was a VFR night for us too. But I had no visual reference in the haze once the lights of towns dotting the East Coast were no longer in my field of view. This lack of visual reference, leading to spatial disorientation, was determined as the probable cause of JFK Jr.'s crash.[65] This was the same result he might have gotten if he barreled into a cloud earlier that day. VFR into IMC (instrument meteorological conditions) is a common cause of accidents among GA pilots. Such accidents are usually fatal—more than 70%. With the loss of visual reference, the untrained will give in to their vestibular system and quickly learn how poorly adapted it is to flight. According to a study by the Air Safety Institute, it takes just under two and a half minutes for the average VFR pilot to lose control in instrument conditions. The data from the JFK Jr. crash is eerily consistent with this average.

---

[65] Spatial disorientation is defined by the FAA as the body misinterpreting the pilot's position in space. You think you're level, but your body tells you you're turning, for example.

From the first unexplained right turn that then transitioned to a descending, tightening left turn before impact with the water was just under three minutes.

Beware of flying at night (and even daytime) with haze. Also beware of taking off at night from a runway with a departure leg that takes you immediately over water. The risk of spatial disorientation in both cases is high.

The stage check

A few weeks after my first night cross-country flight with Connor, I was scheduled to do a "stage check" with another instructor before being turned loose on my first solo cross-country flight. This was a way to ensure that Connor wasn't drinking his own Kool-Aid, a way to get another set of eyes to confirm that I was indeed ready for some alone time on a long flight. I was paired with one of the older instructors at the school. And by older, I mean he might have been five to 10 years older than me. Despite his gray hair, I was less intimidated by him than Connor. He asked me to plan a cross-country flight and said that we'd fly toward my planned destination—I picked an airport to the southwest of Hanscom in Connecticut called Windham (KIJD)—and then, after he was satisfied with my dead-reckoning and pilotage skills, we'd return to Hanscom and do some touch-and-goes.

We met early on the day of the flight and went over my flight plan. He was complimentary but was critical of my choice of the Putnam VOR as a visual checkpoint along the way. He said that VORs are hard to see from the air and that I might want to choose something else. When I originally contemplated using the VOR as a landmark along the route, I too worried that it might be hard to spot. But I had done my homework. On a whim, I had gone to Google Maps and typed "Putnam VOR" into the search bar of the mapping application. And wouldn't you know it? Doing this took me straight to the VOR.[66] Turning on the satellite image overlay showed me that that the station sat on a hill in what looked to be an open field, clearly defined rows of trees demarking its boundaries. I also noticed that it was located almost halfway between Interstate 395 and a body of water called the Quaddick Reservoir. I assumed these large landmarks, along with a moving map on my iPad and tuning to the VOR signal while watching for the needle start to "swing" as I got close, would make it easy to find the station. The instructor was incredulous. He told me that we'd try it my way, but in a tone that clearly indicated that he didn't think my way would work.

I called for flight following on the ground at Hanscom and said I was going to Windham and then phonetically spelled the airport ICAO code: Kilo, India, Juliet, Delta.

"You don't need to say 'Kilo,'" my instructor said. I sensed some annoyance.

"Before ForeFlight and other EFBs came along, nobody used 'Kilo.' You're in the United States. They know it begins with 'Kilo.'"[67]

I felt more chastised than I'm sure he intended. But his comment had a "young whippersnapper thinks he knows what he's doing but he really doesn't" vibe. I imagined him talking to

---

[66] And, by the way, it turns out that searching for a VOR in this way *almost* always works. It helps to be specific in your search; sometimes using "VORTAC" or "VOR DME" is helpful.

[67] A friend of mine who's an airline captain for a major carrier said that he disagrees with this advice. He noted that since some VORs and airports have the same root name, ATC will use the *K* or say "airport" to ensure there's no confusion.

his wife at the end of a long day of flying, complaining about how nobody knew how to use a manual E6B anymore. I kept my opinions on that topic to myself.

My sensitivity to his frustration with the use of the term "Kilo" didn't keep me from immediately frustrating him again. After calling for flight following, I waited for the controller to return with my squawk code. I was conscious of not asking for taxi clearance until I had the code as I didn't want to be maneuvering the airplane on the ground while trying to type in the code on the transponder.

"What are we waiting for?" he asked.

"I was waiting for the squawk code, I …."

"I can type in the code," he said, cutting me off. "Let's get going."

Okay then.

I began my taxi, and the controller came through with the code a few moments later. The instructor typed it in for me.

We taxied out to Runway 29, and I departed to the southwest. He gave me an altitude to fly and asked me to put on my foggles for a while and fly by instruments between my first and second waypoints.[68] Despite what I perceived as his angst with me, I performed well. He quizzed me on various topics as we flew. I did a good job of multitasking, answering his questions correctly while keeping the plane on course.

As we neared the Putnam VOR, I began to scan the land beneath us. About five miles from it, I had it in sight. I pointed it out to the instructor.

---

[68] Foggles are glasses with lenses that are frosted on the upper one-half. This prevents the wearer from seeing out the window of the aircraft but allows them to see the instruments.

"Well, what do you know?" he said, seeming genuinely impressed that I'd been able to locate it from the air.

With that, he asked me to turn the plane around and head back to Hanscom. He contacted ATC and told them what we were doing.

As we flew back to Hanscom, he gave me some instruction. As I noted earlier in the book, spending time with more than one instructor during your training can be very beneficial. He told me that I didn't "dance on the rudder pedals" enough. He was right. In shallow turns while in cruise flight, it's easy to forget about the use of the rudder. While using the rudder while in cruise flight may not have a huge effect on safety or aircraft performance, it's good, as the instructor pointed out, to have your hands tied to your feet so that whenever the yoke is turned your feet do something. That way, when it is more critical (e.g., sharper turns at slower speeds), you are fully coordinated. To this day, I've taken that advice and try to make it a point to use the rudder even with the lightest of pressure during shallow turns at higher speeds.

"Thanks for noticing that," I said, trying to please him after feeling like I had been a pebble in his shoe for the first half of the flight.

"No problem," he said, and then added, "I had to come up with something to be critical of since you seem to be more than prepared."

I beamed on the inside.

While his instruction on the use of rudder pedals was helpful, there was one piece of advice I was more skeptical of. As we entered the Class D airspace around Hanscom, he said that he liked to be at pattern altitude the minute he crossed the imaginary dotted blue line that defined the airspace on a sectional. That would mean being at about 1,200 feet MSL five miles from the airport. His logic was that this made it easier to see other traffic in the pattern and for that traffic to see you, reducing the risk of midair collisions.

I could see the logic, but like a lot of things in aviation, increasing safety in one regard decreases it in another. I didn't like being that low so far from the airport. If throttling the engine up and down closer to the ground happened to lead to some sort of power loss, I'd be looking at putting the plane down in less-than-hospitable topography. The area around Hanscom is, as I mentioned earlier, heavily populated and wooded. The only viable option outside of trying to guide the plane into treetops would be ditching in small bodies of water here and there.[69] If we were flying into an airport in Nebraska with plowed fields as far as the eye could see, I'd understand the logic of his advice. But at a place like KBED, I wasn't a fan. Still, I figured it wasn't worth getting into a debate while low, slow, and close to a busy Class D airport, so I did as he suggested. But it's the first and last time I've done that.

---

[69] The data on survivability of the water vs. trees is mixed. Ditching in water would be the better option based on work done by Paul Bertorelli, a journalist from AVweb. He notes in several articles and videos that based on the data, survival rates for general aviation water "landings" (as the airlines call them during your safety briefing) are about 90%. But a separate study by the staff at *Aviation Safety* summarized in an article called "Water or Trees?" notes that while injuries are less severe in water landings, survivability of controlled ditching into trees is about the same.

On my own

I was finally ready for my first solo cross-country flight. The plan was to fly to KPWM (a Class C airport in Portland, Maine), then to KPSM (a Class D airport in Portsmouth, New Hampshire), and back to Hanscom. It was a nice September day with fair-weather, scattered clouds at 5,000 feet and calm winds. Connor sat down with me prior to my departure to review my flight plan. After I satisfied him that the plan was solid, he took my logbook, turned to the back of it, and entered an endorsement, essentially giving written witness to his belief that I was prepared to fly by myself a total of 165 nm with landings at three airports. After he completed his entry, he handed me my logbook and wished me luck. The FAA requires student pilots to carry their logbooks with them on solo cross-country flights. That's because the endorsement in the logbook is your legal permission to do the flight. If an FAA representative "ramp checked" you at an airport while on the cross-country flight, you'd want to have that documentation with you. I thanked Connor and, with that, I was off.

My landing at Portland was my first at that airport and, in fact, it was the first time I landed at a Class C airport. I didn't find it much different than landing at a Class D with one exception: I was cleared to land behind a Southwest Airlines 737. In all my landings at Class D airports during training, I hadn't encountered large commercial aircraft. I was about 10 miles behind the jet but stayed a bit high in any case just to be extra cautious. Wake turbulence settles like a leaf falling from a tree; you are always safe above the flight path of the larger plane. After a while though, I realized I had sufficient spacing from that traffic and descended to a glide path that would ultimately intersect the "numbers" on the end of the Runway 36. Little did I know that wake turbulence worries would become much more pronounced in just a short while.

Throughout my flight, from the time I contacted ground at Hanscom, to asking for flight following, to landing at Portland, I was sure to insert "student pilot solo" into my first transmission to a given controller, hoping for the grace that comes with being a beginner. The controller at Portland was very helpful. The airport was quiet when I landed. Nobody else was in the pattern and the airliner that had been well ahead of me was now parked at its gate. I had a map of KPWM handy, but after landing I told her that I was unfamiliar with the airport and asked for progressive taxi instructions. She obliged, telling me to turn left on taxiway Alpha then left on taxiway Charlie, approved crossing Runway 11/29, and told me to hold short of Runway 36. I was back in the air in no time and made my way south to my next stop: KPSM.

Portsmouth Airport was, for years, known as Pease Air Force Base. It had a bomber wing stationed there that flew FB-111 Aardvark fighter/bombers. The FB-111 was a sleek-looking airplane with two afterburning engines, a long nose (thus the "Aardvark" moniker), and wings that could be swept forward for slow flight and back for speeds exceeding Mach 2. The airplane had a special place in my heart because my father flew in it during his Air Force career. Not only did he fly in it, he ejected from one on a fall day in 1976, coming close to meeting his maker. Believe it or not, I was thinking of that flight that almost killed my father while I was on my cross-country. There were important lessons in that crash that apply to me … and to you.

***

First, some background. My dad was a weapons system officer (WSO) in two-place fighter/bombers, not a pilot. He began his career and had most of his hours in the back seat of the RF-4C, the tactical reconnaissance version of the famed "Phantom" fighter. In *Top Gun* parlance, he was "Goose," the back-seater. The F-111 was a fun airplane for him though because, while he wasn't the pilot, he also wasn't strictly a "back-seater." In the Aardvark, the WSO sat next to the pilot and therefore had a great view out the front windscreen.

There were many reasons to love the airplane, but it was being in front that made it a real joy for Dad. He was certainly relishing the idea of a front cockpit view on his first F-111 flight on October 27, 1976, from his Florida panhandle base, Eglin AFB. The primary mission for that day called for the pilot to fly the aircraft out over the Gulf of Mexico and do several high-speed passes (all greater than the speed of sound). On each pass, Dad was to open the bomb bay doors for 30 seconds to allow data collection from sensors on a dummy weapon in the belly of the aircraft. The data would then be transmitted to engineers on the ground. Unlike my father, the PIC for the mission that day had significant time in the F-111.

The setup for the test involved positioning the aircraft over the Gulf of Mexico about 60 miles south of Eglin. While flying to the test area, the pilot climbed to 40,000 feet MSL. When it came time to collect the first data point, he swept the wings full aft, ignited the afterburners, and accelerated to Mach 2. Dad opened the bomb bay doors, and the first data collection was complete. The pilot then began a stair-step descent to the lower altitude data points. Significantly, throttles remained in full afterburner to achieve the supersonic speeds required for the test.

In the lower, thicker atmosphere, the fuel-flow rate increased exponentially.

Shortly after the second data point was collected, my father cautioned the pilot they had reached "bingo fuel," a pre-calculated fuel-remaining state that dictated when the mission should be aborted (if it had not been finished) and the jet flown back to base. This ensured enough fuel to make it back to the field and even do a go-around. But the pilot elected to continue.

This was the first critical mistake—a first hole in the Swiss cheese of this story. Had Dad been in an F-4 Phantom, an aircraft in which he had over 1,500 hours of flight time, he would have vigorously challenged the pilot's decision to violate the bingo-fuel call, particularly when sustained afterburner power was being used. But this was his first flight in an F-111. He reluctantly deferred to the pilot's experience.

Shortly after the collection of the last data point, the low-fuel caution light illuminated. This shocked the pilot (and my father). It did not seem possible that so many thousands of pounds of fuel had been consumed in such a brief period. The pilot suspected the fuel gauge might be malfunctioning or there was a fuel leak in the aircraft. As it turned out, there was nothing wrong with the gauge and there was no leak.

As the pilot continued back to Eglin, a low-fuel emergency was declared, and Dad began going through the landing checklist. One of the steps in that checklist was to look up the appropriate wing sweep for landing (which varied with the plane's remaining fuel and was critical to maintain the plane's center of gravity as it slowed to land). The pilot set the wing sweep to the correct position. All seemed good for the approach. However, because of the distraction of the low fuel state and concern there was a fuel leak or malfunctioning gauge, the pilot's landing approach was high and fast (Swiss cheese hole number two). The plane touched down well above landing speed, and immediately bounced into the air.

At this point, Swiss cheese hole number three comes into play. At touchdown, during normal landings, F-111 pilots swept the wings full forward. This resulted in increased drag and shortened the landing roll. Dad's pilot instinctively made that wing sweep change as he normally did.

Unfortunately, this was not a normal landing. The plane was already back in the air after the bounce and the full-forward wing sweep put the aircraft's center-of-gravity out of limits. The plane became uncontrollable. Meanwhile, my father was shouting at the pilot to initiate a go-around, less concerned about a low fuel state than the landing ending badly.

The jet was still flying above landing speed as the pilot repeatedly forced the aircraft to the runway (more Swiss cheese holes), to be followed by ever more violent bounces into the air, resulting in more calls from my father for a go-around. Remaining runway continued to be eaten up, the plane was increasingly out of control, and the pilot was oblivious to repeated go-around pleas. Dad was certain death was imminent. He initiated an ejection. In the F-111, the entire cockpit, with the two crew members inside, is blasted away from the fuselage, a parachute is deployed, and the cockpit capsule lowered to the ground. In this case the ejection was at extremely low altitude; the capsule had one swing on the parachute and slammed onto the runway several hundred feet ahead of the jet's carcass, which caught fire due to the small amount of fuel remaining in the plane. My father and his pilot escaped with minor injuries.

The accident investigation concluded the pilot failed to terminate the mission at bingo fuel and then failed to attempt a go-around after it was apparent the first landing attempt was unsalvageable. He was grounded and retrained. My father later flew with him many times. As he said, "After that near miss, he was the safest pilot I ever flew with."

***

What lessons are there for us in this story? First, if you are flying with another pilot, look after each other. Don't assume you don't have a voice. Give each other permission to disagree if doing so leads to better decisions. Second, don't deviate from known safety principles. If you reach a fuel state you know is only good to get you back home and do a couple of approaches, go home. Don't tempt fate. Finally, and most importantly when it came to my cross-country flight that day, don't forget the axiom "aviate, navigate, communicate … *in that order*." Don't allow yourself to be distracted by navigation and communication (or an assumption about a fuel leak) to the point that you lose control of an aircraft. *Always* aviate first.

Good ADM … I think?

It was easy to tell that Portsmouth was once a military base. Its only runway is over 11,000 feet long. For a Class D airport not handling intercontinental flights, this is long. Boston's longest runway is more than 1,000 feet shorter. Atlanta and Los Angeles each have runways over 12,000 feet in length, but they handle very large, heavy aircraft in places where it can get hot. Periodic high density altitude significantly increases the need for longer runways. I could do five landings in one approach at Portsmouth—like a rock skipping along the water—given the amount of asphalt stretching in front of me. It was that long.

I contacted the tower and was told to enter a right downwind for Runway 34. As soon as I got on the downwind, though, I was told to do a 360° turn for spacing. It wasn't an unusual request. I had been asked many times at Hanscom to perform the maneuver, usually to accommodate faster traffic in or entering the pattern. But I quickly realized this was a different situation. I was being asked to do the turn to accommodate not one, but two KC-46A aircraft, the Air Force air-to-air refueling version of the Boeing 767. The 767 is a wide-body jet used for long-haul routes in the commercial aviation world. The refueling version's max takeoff weight is 415,000 pounds. To put that into perspective, the max takeoff weight of the largest, newest Gulfstream business jet (the G800) is about 108,000 pounds. Tankers are *big.*

The two airplanes were trailing each other and doing touch-and-goes. The controller was trying to find a way to fit me in between them. But as I watched them landing below me, I noticed dust being kicked up all over the field. The wake turbulence of an airplane the size of the KC-46A's is like the wake of a humpback whale in the ocean. The air is roiled into dangerous swirls.

As the second aircraft completed its touch-and-go, the first was already on a downwind leg on the opposite side of the runway from me (a left downwind). The controller started to tell me I could reestablish myself on the right downwind but then said, "November 275 November Delta, disregard. Give me another 360 for spacing." I complied and noticed the tanker turning another base to do yet another touch-and-go. The evidence of air swirling all around the runway was clear even from 1,000 feet up. The combination of touch-and-goes (which meant the air never really had a chance to settle) and the presence of two heavies in the pattern started to worry me. The risk of a mishap from wake turbulence was non-trivial; I didn't want to risk it. I made a command decision.

"Portland Tower, November 275 November Delta," I said. "With all you've got going on, cancel my landing request. I'd like to extend my downwind to the east and then exit the pattern to the south if that works for you."

He seemed relieved to hear of my desire to lighten his load and asked me to fly to the coastline before I turned south. I did as he requested and was back on my way toward Hanscom.

As I flew away, I realized that I now wouldn't have the second of three full-stop landings at three different airports I needed to satisfy my solo cross-country requirement. With this in mind, I made another command decision. I'd divert to Lawrence Municipal Airport (KLWM) 25 miles to the southwest of Portsmouth. I had landed at the Class D airport several times during my training and was therefore familiar with it. It had the added benefit of sitting right between Portsmouth and Hanscom too; I wouldn't have to fly out of my way to get another landing in.

My approach and landing there went beautifully and I was quickly on my way back to Hanscom, where my third landing was also silky smooth. I taxied to the ramp, parked, buttoned up the airplane, and made my way back to the flight school's offices.

"How'd it go?" Connor asked me as I entered the building.

"Great!" I said, feeling the same spring in my step I had after my first solo in the pattern. "Really was a lot of fun."

He asked me about my comfort on the radios and if there were any issues on my approaches to the airports.

"No," I said. "It all went great. I'm feeling very comfortable."

Then, I added almost as an afterthought, "But I did have to divert to Lawrence instead of landing at Portsmouth because of some heavy aircraft in the landing pattern."

His face went white.

"What do you mean?" he asked.

I explained why I had decided to leave the pattern at KPSM and instead go to KLWM. After I was done telling the story, his face was a bit whiter. I could tell that worry was hanging on him.

"What did I do?" I asked, genuinely perplexed.

Readers who are smarter than me probably know what I did. Flash back to Connor filling out the endorsement in the back of my logbook: I had assumed that the endorsement to complete the solo cross-country flight was generic in nature, that it essentially said, "I believe Patrick is capable of doing a cross-country flight with stops at two airports besides his home airport." I never actually read what the endorsement said when Connor had completed it in the back of my logbook. And what it said was this:

> I have reviewed the cross-country planning of *(First name, MI, Last name)* Patrick Mullane and find the planning and preparation to be correct to make the solo flight from *(location)* KBED to *(destination)* KBED via *(route of flight)* KPWM, KPSM with landings at *(name the airports)* KBED, KPWM, KPSM in a *(make and model aircraft)* PA28-161 on *(date)* 9/29/21. *List any applicable conditions or limitations:* ECAC student pilot limitations.

In an area under this information was Connor's signature, the date, an expiration date, and his CFI number. Conspicuously absent from the very explicit instructions in the endorsement are the letters "KLWM"—the code for Lawrence Municipal Airport. I was supposed to have stuck to the airports listed.

Connor explained to me that he had confidence in me and my flying skill and added that he thought not landing at KPSM was the right call. But he was worried that come checkride time, the discrepancy between my logbook entry that said I landed at KLWM and the endorsement that mentioned KPSM would be a problem. I tensed up thinking I had really screwed up before I'd even made it to that checkride.

I wasn't one for fudging things—neither was Connor—so the logbook entry for the flight was going to stay as it was. I landed at KLWM. So be it. Nothing I could do about it now and I never considered "making it go away" by just writing in the logbook that I had landed at KPSM. Connor told me that he would check with a DPE to get their opinion on whether my error was a forgivable one. I thought this was the prudent route too. Better to know now if my actions were problematic (well, problematic enough to warrant some sort of reprimand, retraining, and/or re-flight) than wait until the day of the examination and have somebody make a stink about the discrepancy in "real time."

A week later, relief. Connor told me he had checked with a DPE, and they said that they thought my actions showed great ADM (aeronautical decision making), that it was nothing to worry about. And, indeed, it turned out not to be. But just in case, I completed one more cross-country flight that met my PPL requirements and that matched the information in the endorsement.

The lesson in all of this? I'd encourage you to read the endorsement language whenever one is provided in the back of your logbook. Turns out there are details in there you might find relevant.

Chapter 18
The Weather and Atmosphere

On my first solo cross-country flight up to Portland, the weather was spectacular. Scattered, puffy cumulus clouds at around 5,000 feet MSL sprinkled the sky. The air was the sort of calm you only get on a pleasantly warm, but not hot, fall day. It was clear too. I could see a million miles. The tranquility of it all helped me relax and enjoy my first time on a long trip by myself.

The wonderful weather had me reflecting on how much time I'd spent studying meteorology during my training. It was significant, which I suppose makes sense since the atmosphere is the playground of a pilot. It's hard to appreciate the power and scale of that playground. The atmosphere weighs over five *quadrillion* tons (the ocean only weighs about 270 times what the entire atmosphere does).[70] That's a lot of mass. And all that mass is constantly moving, ultimately driving every meteorological event we experience. What makes it move? A single thing: the uneven heating of the Earth's surface.

---

[70] A quadrillion is one followed by fifteen zeroes: 1,000,000,000,000,000.

Uneven heating at the local level

When a parcel of air is heated, molecules spread apart, making it less dense relative to a cooler parcel of air. Less dense air rises relative to surrounding cooler air—this is the reason a hot air balloon climbs. This results in movement that manifests itself in two important ways at a local level: wind and convection. You'll deal with these two things every day as a pilot. The direction and speed of wind must be considered when landing and navigating—even when taxiing. It's ever-present and usually a nuisance unless it's pushing you along from behind on a long cross-country flight.

By use of the word "nuisance" I don't mean to imply that wind isn't dangerous. A 2010 FAA weather-related aviation accident study noted that wind was the leading weather-related cause or contributing factor in general aviation accidents. That's not surprising. Even on a day that seems dead calm when you're doing a preflight on the tarmac, there are winds aloft to contend with. I suspect the fact that it is ubiquitous leads to a complacency among some pilots that can get them into trouble. Thus, the accident data.

If wind is the horizontal movement of air, then thermals are examples of the vertical movement of air. And the uneven heating of the Earth is the origin of both. For example, an onshore wind at your local beach is the result of the land heating more quickly than the water in the daytime, causing the air over land to rise. This allows cooler air over the ocean to rush in and fill the void, explaining the sea breeze you experience when sunbathing at the beach on a warm summer day. Depending on the time of year and the local geography, this process can reverse at night when the land cools faster than the water, making the air over the ocean warmer relative to that over land. The cool air over the land flows to "backfill" the now warmer air that rises over the ocean. This cycle of warm air rising and being replaced by cooler air explains almost everything in meteorology. And it happens on a scale much greater than that of your local beach. We'll get to that shortly.

Another localized effect of uneven heating is the formation of thermals. Thermals are caused by convection—the rising of warm air. As a kid flying remote-control gliders when growing up in southeast Texas, I learned quickly that launching my glider in the Texas summertime heat from a grass field surrounded by housing developments was a guaranteed way to gain altitude. The asphalt roofs of the homes would get piping hot—much hotter than the grass of the field I was launching from. As soon as my glider crossed the boundary of the field and entered the air over the rooftops, it climbed as if on an invisible elevator. I'd enter a turn as soon as I sensed the increase in vertical speed, trying to keep the model plane within the vortex of rotating, rising air and continue the climb. Flying in and out of thermals in a powered aircraft is what causes the ups and downs—the "bumps"—on a warm day. While prone to make a ride uncomfortable, particularly for nervous passengers, convection is not generally a problem for aircraft, with one *enormous* caveat: convection is a key ingredient in the formation of thunderstorms.

I'm a huge fan of thunderstorms. I've always loved to watch them form and experience their majesty and violence both from afar and from underneath them. There's a sort of beauty in their ferocity. But don't let their beauty fool you; these storms are violent. Very violent. They can literally rip a plane apart. In 2011, a pilot in Texas flew his light, single-engine aircraft too close to a thunderstorm (a relatively weak one based on radar imagery) while trying to navigate around a line of weather, apparently unwilling to land or divert, hell-bent on making it to his destination.[71] It was a bad decision. The plane ventured into the edges of the storm and the associated turbulence. The left wing was ripped from the fuselage. The plane and its five passengers, including two children, plunged several thousand feet to the central Texas ground. It's a horrible thing to ponder but this accident and many others are how we remind ourselves of what we've been taught: thunderstorms are not to be toyed with.

Thunderstorms form when three things are present: a lifting force (i.e., convection), moisture, and an unstable atmosphere. Convection is the elevator that lifts air to build the cauliflower cumulus clouds that are the harbinger of thunderstorms. It may be obvious that moisture is necessary for a thunderstorm to form—after all, rain is … well … moist. But there's a more important reasons moisture matters, and it has to do with the last ingredient that goes into our thunderstorm stew: unstable air. What is unstable air and why do we need it?

Stable air vs. unstable air

Before getting into the details of stability and instability, one quick note. You may already know that as altitude increases, temperature decreases. In fact, there's a rule of thumb that says that for every 1,000 feet in altitude gained, the temperature drops 2° C or about 3.5° F. This rate of

---

[71] The pilot also had in-cockpit radar information, but it was delayed, not real-time. He may have believed he was farther from danger than he really was.

temperature change based on altitude is known as the "lapse rate." It's one reason flying an unairconditioned airplane in the summer can be bearable. A day with temperatures around 90° F on the tarmac can give way to a comfy 72° F or so at 5,000 feet. That lapse rate is important to understand in our discussion of stable vs. unstable airmasses.

In both stable and unstable atmospheres, warm air will rise. And as a parcel of air rises it will expand because the pressure around it will decrease. This expansion results in cooling. In stable air, this cooling decreases the temperature of the parcel to a point that it matches the temperature of the air around it at relatively low altitudes. This has the effect of keeping the air from rising farther, killing the lifting force a thunderstorm needs to form.

In unstable air, this lifting force persists. But why? There are several forces at work. A lapse rate where temperatures cool more quickly over a given altitude can be a factor. This makes it likely that a warm parcel remains warmer than the air around it even as it rises and cools, allowing it to ascend farther into the atmosphere than it otherwise might. Also, as I alluded to earlier, moisture plays a part. Moisture in a rising, cooling parcel of air will eventually condense because of that cooling, much like the moisture in the air condenses on the side of a cold glass filled with your favorite beverage. This condensation is visible and, in our atmospheric example, is what you see as a cloud. But here's the interesting thing about the process of condensation: it generates heat.[72] This heating inside the parcel

---

[72] Condensation generates heat because the relative movement of atomic particles in gasses is greater than that in liquids. As the molecules organize themselves into a liquid from a gas during condensation,

of rising air mitigates the cooling that's occurring during its expansion as it treks upward and keeps the parcel warm relative to the air around it. This results in even *further* lifting. (Note that the relative temperature is what matters here. A parcel will still rise even if it's 35° F if the air around it is even cooler.) This express elevator to tens of thousands of feet is critical to thunderstorm formation, and it's the reason warnings about convective activity should never be ignored.

These details aren't ones you'll be considering when flying. But my hope is that the mental image of the mechanisms that cause thunderstorms will keep you on your guard during all phases of flight. Whenever you hear of a warm and wet atmosphere being unstable, keep an eye to the sky. It will be obvious what to look for. That's because while the elevator of rising air that starts the process of thunderstorm development is not directly visible, the result of it is: cumulus clouds billowing and building like cotton candy on an invisible thermal stick. These beautiful clouds may not be dangerous yet, but as they grow and reach for the stratosphere, they can become so, and *very* quickly. As they start to flatten out at their tops, beginning to form the anvil head that makes a thunderstorm so distinctive, they are to be respected. The FAA recommends that you stay at least 20 nm away from a thunderstorm; the physical form of the thunderhead doesn't define its zone of impact. It can throw hail into clear sky; it can produce unseen

---

conservation of energy requires that that decrease in kinetic movement must result in an increase in some other form of energy. That other form is heat.

winds and turbulence miles from its center and outside of its white froth. Just stay away.

Pressure altitude – the path to density altitude

One advantage of thunderstorms when it comes to flying safely is you can typically see them when you are flying in VFR conditions. That makes it easier to avoid the dangers they present. Not so when it comes to density altitude. You can think of density altitude as a "feels like" altitude for an airplane. By that I mean it's a variation on the "feels like" temperature your local television meteorologist uses to explain that while the temperature outside is 30° F, it feels like 20° F because of a stiff wind, or 80° F feels like 90° F because of brutal humidity. Density altitude is your plane's "feels like" altitude because it describes what the air feels like to the airplane *regardless* of its true altitude above sea level. And the altitude an airplane "perceives" matters more than anything else when it comes to its performance.

How is density altitude determined? Let's start with the formal definition of what it is: density altitude is pressure altitude corrected for nonstandard temperature. Gotta love a definition that requires another definition. Sounds like we need to know what pressure altitude is. Let's start there.

First, remember that pressure altitude can be easily determined by setting the pressure in the Kollsman window of your altimeter to a standard pressure of 29.92" or 1013 millibars of mercury and then reading what the altimeter says. But understanding why this works is important to understanding how an altimeter functions and, ultimately, understanding density altitude.

Remember that an altimeter is just a sensitive barometer. It doesn't really measure altitude; it measures pressure. The pressure measured drives the "hands" of the altimeter (at least in a traditional analog instrument), displaying how high you are above sea level (if it's configured properly). Imagine for a moment an altimeter in a world where there were no high- and low-pressure weather systems, where the *only* thing that changed the pressure the altimeter "felt" was an increase or decrease in altitude. In this world, the altimeter could be hardwired; an altimeter at sea level would always read zero and one at the top of Pikes Peak in Colorado would read 14,115 feet. Every. Single. Day.

But of course, we don't live in a world with constant atmospheric pressure. It varies constantly. For example, as I wrote this sentence, a graph of the last 72 hours of barometric pressure readings in Denver, Colorado, looks like a Coney Island roller coaster. The graph dips and ascends several times, peaking at 30.68" Hg and registering a low of 30.05" Hg.[73] This variability creates problems for a hardwired altimeter. It will be faked out into thinking that every change in pressure was altitude related when it may not have been.

To illustrate the point, suppose you were working a shift in Allison's Amazing Altimeter Factory in Denver, Colorado. You are at the end of the line. Your job is to set the dials on the face of the instrument to your altitude—5,280 feet—and then attach those dials to the tiny barometer in the back of the unit. You are finishing up your last batch of altimeters when the end-

---

[73] Hg is the chemical symbol for mercury. Mercury is known as "liquid silver," an apt name if you've ever seen it poured onto a table. Liquid silver in Latin is hydrargyrum—thus our Hg abbreviation.

of-shift whistle blows. It's time to hit happy hour with your friends. You leave the shiny, new units on your workbench and head out for the bar, oblivious to the fact that the air pressure was 30.05" Hg when you set the altimeters to your mile-high altitude. The next day, you come back to work. Still oblivious, and maybe a little blurry-eyed from the happy hour the night before, you take your place at your workstation, not realizing that the pressure has climbed overnight to 30.68" Hg. You look once again at the altimeters you had set the night before. Do you think they will still read 5,280 feet?

At the end of your shift, you set the altimeter to 5,280 ft. The air pressure is 30.05" Hg.

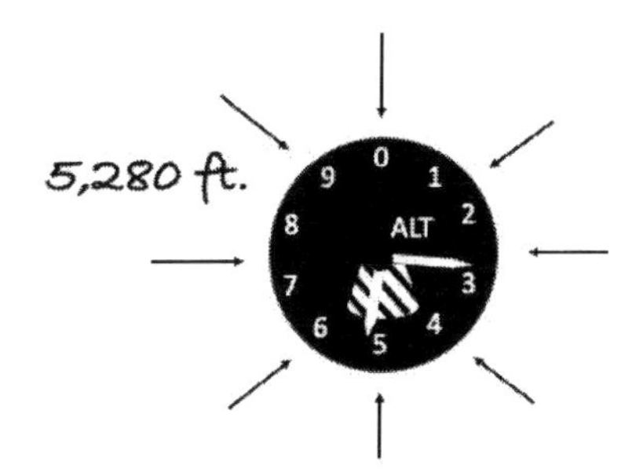

No, they won't, even though they are still the same distance above sea level they were the night before. They will read *lower* than the altitude you set at the end of your previous shift. Why? Since your hardwired altimeters think that a change in pressure can only be because of a change in altitude, it interprets that higher pressure as evidence that there's more atmospheric weight above it; it thinks it has descended.

The next day, the pressure is up to 30.68" Hg. The air pressing on the altimeter is "heavier." It thinks this is because it is lower – i.e., more of the atmosphere is above it.

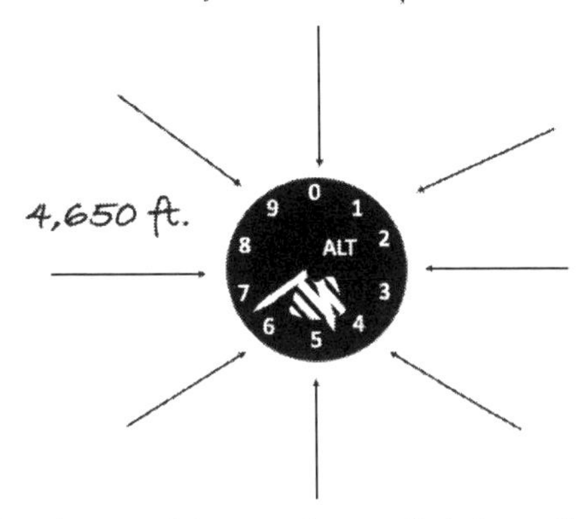

This is why an altimeter has a dial that allows you to input an altimeter setting that shows up in the Kollsman window. By spinning the dial to select the altimeter setting at your position, you are correcting the altimeter's reading for any pressure difference not related to a change in altitude—you are correcting it for variations in the atmospheric pressure.[74] It's as

[74] Contrary to common belief, the altimeter setting is not the same as the "raw" barometric pressure reading at a location unless you are at or very, very close to sea level. Rather, it is barometric pressure corrected to sea level pressure, meaning it's an adjustment to say, "If this air mass was at sea level, what would the pressure be?" This correction is made to be able to do an apples-to-apples comparison of pressures across locations. To illustrate this point, consider that the lowest pressure ever recorded was in the eye of super typhoon "Tip" while it was off the coast of Guam, coming in at 25.69" Hg. This isn't surprising since typhoons and hurricanes are just enormously powerful low-pressure systems. As of the writing of this footnote, though, the uncorrected pressure at Denver International Airport is 24.86" of mercury. That seems to imply that the pressure in Denver is consistent with a *super-duper* typhoon sitting on top of it. A sign of the end of the world if true. But Denver's lower "raw" pressure isn't because of a

if you are resetting it to its factory condition. Rotating the knob until 30.68" Hg shows in the Kollsman window spins the hands of the altimeter clockwise until they settle on your altitude above sea level once again.

Fortunately, this adjustment is easy to design into an altimeter because 1" of mercury equates to 1,000 feet of altitude. Spinning the knob on the altimeter to select a pressure in the Kollsman window is simply engaging a mechanism that adds or subtracts 100 feet of altitude for every 0.1" of mercury. In our factory example where you set your last batch of altimeters to your known altitude on an evening when the pressure was lower than the next morning, the difference in pressure between that end of shift reading and the morning whistle was 0.63" Hg (30.68 - 30.05 = 0.63). Your altimeters will read about 4,650 feet, 630 feet lower in the morning than they were the night before. If you turn the dial in the Kollsman window, that 630 feet will be added back to the altimeter—the

---

low-pressure system, it's because of its high altitude. When the pressure is corrected to an equivalent sea level pressure, the reading is 30.28" Hg, a nice high-pressure day.

hands will spin in the clockwise direction to get you back to your actual altitude of 5,280 feet.

Knowing this math is helpful when you are planning a flight and not in the cockpit (because in the cockpit you can use the method I mentioned at the beginning of this section and will talk about more in a moment—just set your altimeter to 29.92" Hg. and see what the altimeter reads to get pressure altitude). To determine your pressure altitude mathematically, first determine the barometric pressure at the time and location of your flight. You can use online resources or electronic flight bags to find predicted barometric pressure; I've found these predictions to be highly accurate. Then, subtract the predicted pressure around the time of your flight from 29.92, multiply that result by 1,000, and add the field elevation. Here's an example:

Predicted Pressure: 30.10

Airport elevation: 1,200 ft. MSL

First step: 29.92 - 30.10 = -0.18

Second step: 1,000 x -0.18 = -180

Third step: 1,200 + (-180) = 1,020

PRESSURE ALTITUDE = 1,020 ft. MSL

Be sure to get the last step right. I like to always subtract predicted pressure from 29.92; that way I know if the result of that step is negative, the pressure altitude will be *less* than the field elevation.

This math is helpful, but you may be wondering why we subtract our barometric pressure from 29.92" Hg to begin with? What's so special about this number? It's only special because it's deemed to be the standard pressure at sea level under very particular conditions. Your altimeter is calibrated so that if it is at sea level under those conditions, it will read 0 feet altitude. As the altimeter gains altitude in a world where atmospheric conditions are always standard, then having your altimeter setting fixed at 29.92" Hg means that if your altimeter reads 5,300 feet, you should expect to be 5,300 feet MSL. In fact, you can calculate the pressure the altimeter must be sensing when it reads 5,300 feet on a standard day by doing the math we talked about earlier. Since 5,300 feet corresponds to 5.3" Hg, then you'd expect pressure at this altitude to be 24.62" Hg (29.92 – 5.30 = 24.62). And indeed, if you check some standard pressures tables, this is roughly the standard pressure at 5,300 feet.

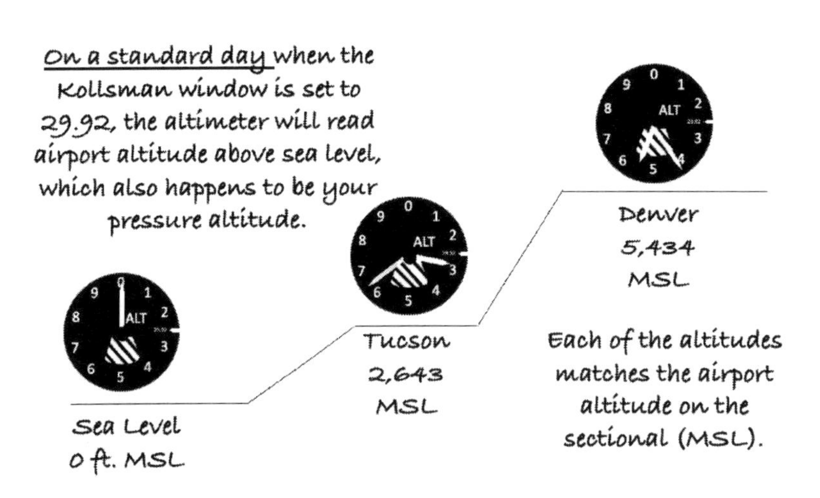

But on a non-standard day, 29.92" Hg won't be at sea level. A barometer might need to be higher or lower than that to sense 29.92" Hg. The line in the atmosphere where it does sense this pressure is called the *standard datum plane.* That line floats as atmospheric conditions change. Sometimes, it might be above sea level; sometimes it might even be below.[75] Imagine a moment when the standard datum plane is at 500 feet MSL. Imagine also that you are at an airport that is at an altitude of 500 feet MSL. With 29.92" Hg in the Kollsman window, your altimeter will read 0 feet. It "thinks" that's where it is because it associates that pressure with sea level under standard conditions. Your pressure altitude in this case is 0 feet. If you then took the altimeter up another 250 feet, it would read 250 feet. But you aren't 250 feet above sea level, you are 250 feet *above the standard datum plane.* This is what we mean by pressure altitude: your altitude above the standard datum plane.

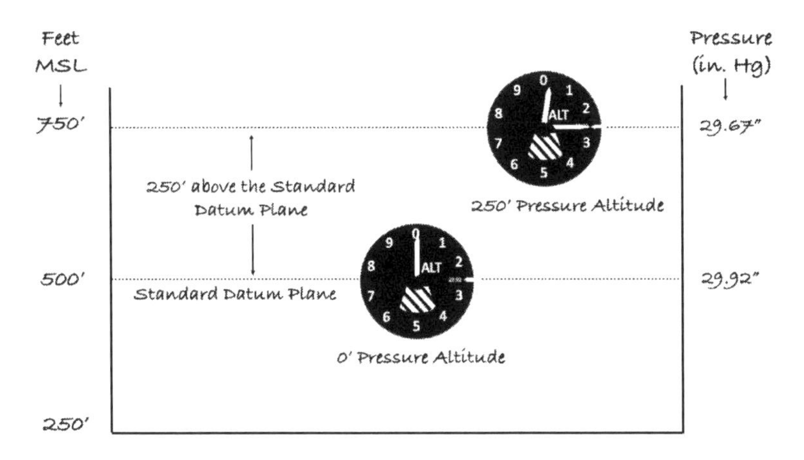

That's all well and good, you may think. But how in the world is that number helpful to us? It's helpful because pressure altitude gets us one step closer to understanding how thick the air is, which has a bearing on how well the airplane performs. That's why you'll find pressure altitude as an input

[75] Obviously, a barometer can't be dunked under the sea to measure air pressure. The altitude at which the 29.92" Hg exists can be calculated mathematically and so when below sea level, is theoretical in nature.

on all sorts of performance tables in your POH. But pressure altitude by itself doesn't tell the whole story.

Density altitude – a short hop from pressure altitude

While pressure altitude shows up as an input in performance tables and graphs in an aircraft's POHs, it is never the only input. It's always coupled with temperature. This is because pressure altitude is anchored off a world of standard pressure and standard temperature, where 29.92" Hg is always the pressure and 15° C is always the temperature at sea level. It's a world where unicorns prance and butterflies flit under a continuously bright, warm sun. In other words, pressure altitude is *not* anchored in reality. To introduce that reality, variations in temperature from those standards must be considered. While our Kollsman window allows us to adjust our altimeter to account for nonstandard pressure, there's no such dial for nonstandard temperature.

Why is temperature important? Because it affects the density of air. And, as I've noted, air density is everything to an airplane.

The concept behind density altitude is straightforward. While it may seem unorthodox to start with a mathematical formula when trying to explain something I contend is straightforward, let's do that here as I think the formula helps us get clarity around what density altitude is:

$$DA = PA + CF\,(Ta - Ts)$$

Now, that's not so bad, is it? It seems even less bad when I tell you what the variables in the formula stand for:

DA = Density Altitude

PA = Pressure Altitude

CF = Correction Factor

Ta = Actual Temperature

Ts = Standard Temperature

The first two of these terms you are by now very familiar with. Let's jump to the last two and then we'll come back to the correction factor.

Since pressure altitude is grounded in a standard day, it makes assumptions about what the temperature is at various altitudes. But as we've already established, the world rarely adheres to standards humans assign to it. That's unfortunate, because if it did, then our density altitude would always be the same as pressure altitude. This should be obvious if we look back at our density altitude formula. If the actual temperature and standard temperature were the same, then $T_a - T_s$ would always be zero and DA = PA. That would be convenient.

But the formula exists because the two temperatures are rarely the same. And since one temperature is the standard and the other is the actual, it's intuitive that you'd want to know how far from the standard you are, thus the need to subtract one from the other. But you may wonder what "CF" is and why we need it. The correction factor takes a temperature difference and turns it into an altitude adjustment. This is convenient since we are interested in calculating an altitude. Well, sort of. As I've noted already, density altitude isn't an actual altitude in that it doesn't tell you how high you are above anything. True altitude (your height above mean sea level), pressure altitude (your height above the standard datum plane), and absolute altitude (your height above land) all do. Density altitude, on the other hand, is theoretical, just like that "feels like temperature" I mentioned earlier. But because an air mass of a certain density mimics the density of air at a given altitude, as pilots we think of density in terms of feet of altitude.

The correction factor is what gets us to that theoretical altitude based on a temperature difference. The number we use for the CF term is 120 ft./°C. For those who have some mathematics background, you can probably see that this figure means that a 1° C difference between standard and actual temperature changes the density altitude by 120 feet.

Let's try an example. I'm currently looking at the weather conditions in Miami, Florida. It's winter and a very strong mass of cold air has descended across the entire United States. The actual temperature in Miami is 8° C (46° F). Not a thong-at-the-beach kind of day. Miami's altitude is just 9 feet MSL—let's call that sea level. You'll recall that standard temperature at sea level is 15° C. Therefore, the difference between standard and actual ($T_a - T_s$) is -7° C. The current barometric pressure is 30.27" Hg. Remember that knowing this, we can calculate our pressure altitude by subtracting the current pressure from 29.92" Hg, multiplying that result by 1,000, and then adding that number to airport altitude. Doing this math yields -350 feet (again, let's just assume that the airport is at 0 feet MSL). So, our formula for density altitude looks like this:

$$DA = -350 \text{ ft.} + 120 \text{ ft./°C} (8°C - 15°C)$$
$$DA = -1{,}190 \text{ ft.}$$

The math may be precise, but the result is just an estimate of density altitude because of some simplifying assumptions. For example, CF is closer to 118.8 ft./°C. Also, the formula doesn't consider humidity. The higher the humidity, the higher the density altitude. This is because humid air is less dense than dry air.[76] But as with some other calculations in aviation, this is directionally good enough for our purpose of understanding how density altitude impacts aircraft performance.

The density of the air in Miami with the conditions as laid out in the previous pages is the density you'd expect if, on a standard day, in a world where oceans had all disappeared, you walked into a dry Atlantic Ocean basin and kept walking downhill until you were 1,190 feet below mean sea level.

If you're looking for better aircraft performance, seeing a negative density altitude is a good thing. Denser air is an airplane's friend. The engine and the wing perform better. Thicker air means you need less runway to get airborne and that the propeller's bites into the air have more "grip." The engine has ample air to feed combustion to develop power. The less dense air of a higher density altitude day works in the opposite direction, degrading aircraft performance. High density altitude will be most noticeable during takeoff. The wing needs a certain number of air molecules flowing over and under it to produce lift. If the air is thick, then when you run

---

[76] The reason for this is that a water molecule (H_2O) is light. You may recall that nitrogen and oxygen make up almost all our atmosphere. But these gasses don't exist as individual atoms. Each typically bonds with a sister atom to form two-atom molecules: O_2 and N_2. Water has only one oxygen atom per molecule. The other two atoms—both hydrogen—are the lightest of all elements. So, H_2O is lighter than both O_2 and N_2. These lighter molecules "crowd out" the heavier ones, making moist air less dense.

up the throttle on your takeoff run that magical number of molecules will quickly be flowing over the wing, and you are airborne in a relatively short distance. But if the air is thin, you'll need more runway and a higher ground speed to get the same amount of air molecules flowing around the wing. This explains why runways at airports located at high elevations—and especially those at high altitudes that also experience high temperatures—are very long. The two runways at Bogota, Colombia, at an altitude of 8,360 feet, are over 12,000 feet long. Denver International, which is lower in altitude but experiences much hotter days (the average temperature in Bogota in July is 67° F; in Denver it's 92° F) has a runway that's 16,000 feet long.

Speaking of Bogota, I'm currently looking at the meteorological aerodrome report (METAR) for El Dorado International airport in Bogota, Colombia (SKBO).[77] The temperature is 9° C (48° F). This may seem cool, but the standard temperature at airport altitude is around -1° C, meaning the actual temperature is about 10° C warmer than standard. The barometric pressure is 30.30" Hg. This puts pressure altitude at about 7,980 feet. Using our density altitude formula, this means DA is about 9,180 feet at the current temperature. The density altitude reported in the METAR, which considers some of the details we don't (humidity, for example) shows it at a more precise 9,327 feet.

As I noted earlier, understanding the math that gets you from pressure altitude to density altitude will help some of you more deeply understand the concept of density altitude. But the good news is performance tables in the POH of your

[77] METARs are reports of weather conditions at an airport, updated every hour.

aircraft make things easy on you by providing density altitude numbers if you know the pressure altitude and air temperature. So don't get too worried if this discussion felt a bit weighty. But do know how to use the tables and be sure you understand the impact of density altitude on your aircraft.

Why these topics?

You will spend a lot of time studying several weather-related topics during your PPL training. It's all relevant, but it's my opinion the topics I covered here are the most critical. The danger of thunderstorms—even ones very far away—and the invisible nature of density altitude can lull pilots into the belief that a dangerous day is safe. The good news is that you don't need to know a ton about weather *if* you are as conservative as you should be every time you plan a flight. If you see a thunderstorm, or one forming, stay away. If you are doing takeoff distance calculations and the math shows you'll have 75 feet of margin on your takeoff roll, don't begin that takeoff. If you see a cloud, don't fly into it. If the winds are a direct 35 knot crosswind at your destination airport, don't land there. If ceilings are at 1,500 feet the entire route of your cross-country flight (and you don't have an instrument rating), don't fly that day. You don't need to know the difference between advection fog, radiation fog, upslope fog, steam fog, and ice fog (and what causes each) to know that you shouldn't fly into fog. Don't assume the weather over your home airport is the weather you'll have along your entire route.

Even if you are conservative, you'll learn quickly that the weather apparently doesn't read the forecasts. Early on in my training, I was surprised by how inaccurate those forecasts could be. We humans have come to believe that our smarts and technology allow us to do anything. Put somebody on the moon? Check. Transplant a heart? Check. Have a computer understand our speech? Check. Split the atom? Check. But we still can't, with great certainty and consistency anyway, say what the weather will be a day or two from now. This is critical to remember. You will invariably launch yourself into the great blue only to find it turn into the great grey an hour later, ceilings creeping lower like a suffocating blanket. Always assume this will happen. Thinking about what can go wrong helps you anticipate and plan for an "out." And, if the possible issues begin to pile up or the forecasts look marginal at best, a good pilot's mantra should echo in your brain: takeoffs are optional, landings are not.

Chapter 19
The Checkride

In late September, eight months after my discovery flight, Connor said, almost matter-of-factly, "We need to get your checkride scheduled."

Come again? I thought.

Like many student pilots, I had this strange contradiction simmering inside me. After months of training, of doing boring maneuvers repeatedly, I was desperate to get to my checkride and have the whole thing over with. I longed to be able to schedule a plane without an instructor and go somewhere with my wife for fun. But the minute Connor suggested that I might be ready for that checkride, doubts consumed me.

"Really?" I said to Connor.

"Yeah, I think you're about ready," he said.

Earlier in the month, Connor had let me know that he had a job offer coming from a regional carrier, a feeder to one of the major airlines. It became clear to me in the moments after he suggested we schedule my time with a DPE that he wanted to get me through the checkride while he was still around. While this initially made me worry that he was trying to push things based on his own timeline, I quickly dismissed my own paranoia as ridiculous. I'd been at this long enough. And the idea of having to change instructors again wasn't appealing to me any more than Connor leaving me in a lurch was appealing to him. He was right; I was ready. I got on the DPE's calendar for the second week in November.

We spent the next five weeks putting the final touches on maneuvers in the *ACS*. I also got a few solo flights in, mostly laps in the pattern to refine my approach and landing skills. I rehearsed all that would happen on checkride day. I also scheduled a readiness review with another instructor. Like my cross-country stage check, this was an opportunity for a third party to check my proficiency.

I was paired with Jim for my readiness review. Jim is unusual in the world of flight instruction. He has a gazillion hours and is an instrument-rated pilot and instructor. But he's accumulated those hours only through instruction. He's never flown for the airlines or in any other commercial capacity. He's made his living simply by teaching others. This showed in my time with him; he was a very good teacher.

We flew to the practice area and went through the maneuvers we knew I'd be asked to do during my checkride with the DPE. Of these, the ones that made me most nervous were steep turns and emergency procedures. My nerves centered around precise execution of these maneuvers, not around the more extreme attitudes these maneuvers required (i.e., steep turns require 45° of bank and an emergency descent requires a rapid loss of altitude). But I executed them well and we headed back to the airport to practice short-field and normal landings and takeoffs.

Jim was the instructor I mentioned earlier who showed me how very precise power settings and the timely use of flaps were hugely helpful in a standard pattern. This was a case where some well-done teaching very late in my student pilot training paid significant dividends. In defense of Jeff and Connor though, it was so incredibly rare to get a standard pattern at KBED because of how busy it was; I was often doing 360° turns in the pattern or extending downwinds to accommodate other traffic. We rarely had the chance to consistently practice this power setting / flap setting regime. And time was always short during two-hour training blocks, so flying to another, sleepier airport to get practice in was a chore. Jim and I were lucky on the day we flew; the pattern was quiet, so practicing a standard pattern was doable. While I was thankful for the opportunity to practice approaches in a more regimented way, since getting my PPL I've learned that flying nonstandard patterns during my training was excellent preparation in another way. It wasn't obvious at the time, but I had become adept at judging my distance from the runway when sent on long downwind legs or asked to enter a base directly when returning from a cross-country. I had gotten a feel for the aircraft and how to manage its power and flaps to get myself on a stabilized approach. If you are at a sleepy airport where you always get to do a standard pattern, I'd recommend making it a point to practice energy management when the pattern looks nothing like what you see in neat illustrations of a standard rectangular circuit.

In any case, better late than never. I practiced these power and flap settings on my home simulator and found that the virtual aircraft performed just as the real one did. It was wonderful, last-minute instruction.

After we landed, Jim said that he agreed with Connor; I was ready. I went to bed that night excited to finally be at the checkride stage but prepared to let the nerves of the flight remain with me a while. The checkride was scheduled for November 18.

I was sure, given the vagaries of the weather in New England in the late fall, I'd have to reschedule several times.

"Are you ready to go fly?"

But Thursday, November 18 came around and it was inexplicably beautiful. The high for the day was going to be 70° F; the weekend temperatures would be a much more seasonal 45° F. I knew even before I looked at aviation forecasts that we weren't going to get away with such warm temperatures that time of year without a brisk southerly wind. And sure enough, the wind was going to be out of the south at around 12-18 knots. Not ideal checkride conditions. But since the direction of the wind was going to be almost straight out of the south on a 170°-180° heading, I figured Runway 23 would be used instead of Runway 29. Using 29 would mean that if the wind was at 180° and 18 knots, the crosswind component would be 17 knots, which happens to be the maximum demonstrated crosswind of the Warrior III we would be flying. More worrisome, there would be a tailwind component of about six knots. If we used Runway 23 instead, the crosswind component would be about 14 knots and we'd have a headwind component of around 11-12 knots. Much more desirable.

My DPE was the most experienced pilot I would fly with before or since my checkride. His relatively small size and folksy way of speaking put me at ease immediately. He had a small office within the East Coast Aero Club footprint big enough for a desk, him, and a student pilot. It had a window that looked back into the rest of the office, and that window was festooned with stickers and pictures that made it clear a man passionate about aviation lived on the other side of the glass. Many of the mementos posted there helped put me even further at ease. I chuckled at a cartoon that showed a mother talking to a son. Written above the image of the two was the following:

Son: "Mommy, when I grow up, I want to be a pilot!"

Mother: "I'm sorry son, but you can't do both."

He made himself even more likeable and therefore less intimidating whenever he asked questions in the form of a scenario. His go-to protagonist in such questions was "Joey-Bag-of-Donuts."

"Joey-Bag-of-Donuts and his girlfriend need to get to a wedding in Poughkeepsie," he would say, "and you've offered to fly them to it. The weather is forecast to be marginal VFR at your departure time but is expected to improve throughout the day."

Then, the question: "How would you think about planning such a trip? Would you do the trip at all?"

The gouge ...

I had an idea of what to expect before I ever met the DPE, thanks to "gouge." The term "gouge" comes from military aviation culture and refers to helpful information passed through informal channels. In various online sources, "gouge" has also been defined as "the skinny," "the lowdown," and "the poop." Gouge for me took the form of written summaries from previous examinees and conversations that happened as I crossed paths with those who had already been through what I was about to go through. (On an internet message board discussing the term "gouge," it is helpfully pointed out that the plural of the word is "gouge" not "gouges.")

You may wonder, as I did, where the term comes from. I poked around for a while and found no definitive answer to this question. There was a Mr. Arthur Gouge, an engineer and designer who developed what became known as the "Gouge flap," a design whereby the flap was integrated seamlessly into the wing. But this seems to be a coincidence. There's no record of Arthur having anything to do with starting informal aviation communication networks. The best origin story I found regarding use of the word in this fashion was not a flattering one. Some surmise that since a gouge in the aviation world is often used in the context of tips and tricks related to passing some sort of exam, the term derives from the same place "gouging" a customer does—meaning to cheat them. I must admit that when I first heard of gouge and saw some of the written summaries previous pilots had penned related to their time with my DPE, my first reaction was to wonder whether this was kosher or not. Was this like getting an answer key to a test in advance?

No, it's not. And that's because the FAA itself tells you what you're going to be asked to do *and* gives you the answers in advance. The *ACS*, *The Pilot's Handbook of Aeronautical Knowledge*, the *FAR/AIM*, and the *Airplane Flying Handbook* contain the answers to any questions you'll be asked. Your aircraft POH has all the information you'll need to know about procedures related to the aircraft you'll fly. DPEs will allow you to reference these books during the oral portion of the exam, if sparingly. You're also told that areas of weakness on the written exam will likely be "explored" by your DPE during the oral part of your checkride. You know what those areas are: they are printed on the test report you're asked to bring to your exam. Finally, the DPE is clear about what they are going to ask you to do. Mine told me in advance the airport I should plan a cross-country trip to. He said that shortly into our trip toward that destination, he'd have me divert to another location and would ask me to use my foggles to fly by instruments to that location.

That's not to say that the gouge didn't provide some useful intel. For example, I knew that my DPE used Google Maps to pull up satellite images of the airport and quiz examinees on runway and taxiway markings. I spent time leading up to my checkride reviewing bird's-eye views of Hanscom Field and several other airports to ensure I knew every marking. As I noted before, I've continued poking around satellite images of airports since, usually to become familiar with a destination I've never been to. The gouge on my DPE also said he was a stickler for clearing turns before maneuvers and to always be looking out the windows, scanning with a "see and avoid" mindset.

Flight sims aren't always "games"

The day prior to my planned flight with my examiner, I used my flight simulator to practice the entire checkride, multiple times. You should too if you have a simulator. It's an amazing way to realistically internalize procedures. To the extent such practice helps to get your brain on autopilot (pun intended) on checkride day, you'll be in a better place. Minimizing surprises is what flying is all about. Using a simulator to help in this regard is invaluable.

I started with my virtual plane parked on the west ramp at Hanscom, the engine off. I strapped my checklist to my kneeboard and went through every step in every section, verbalizing as I went. I had my sectional map handy and programed the GPS and nav radios on the ground in my digital plane just as I would the next day in the real airplane, assuming the weather allowed us to fly. I practiced maneuvers with engine settings and altitudes I knew I'd use. I practiced short-field, soft-field, and normal landings and takeoffs. I did everything I possibly could to ensure that nothing was a surprise. This is an important lesson not just for a checkride but also for your life as a private pilot. There are many resources available that can help you plan and practice before you ever climb into a cockpit.

The government always wants its paperwork

The evening before the checkride, after completing my simulator practice, I laid off the studying to allow my brain to rest. I did do one more preparatory thing, though, and that was make sure I had all my physical and online paperwork buttoned up. If you are getting your certificate in the United States and are already a student pilot, you've become somewhat familiar with the FAA's IACRA website. You had to put your information into the IACRA system to get your student pilot certificate. IACRA stands for "Integrated Certification and Rating Application" and is the repository for all information the government will need to know about you on the path to awarding you a private pilot certificate and any certs that may follow. The *New User Guide*, an online guide for IACRA, describes the system this way:

> *IACRA is an FAA web site that allows people to apply for new Airman Certificates, or to upgrade their existing certificates. The entire application process takes place on the website, including electronic signing of applications, and at the end of the process all the necessary documents will be sent electronically to the Airman Registry.*

As you get ready for your checkride, be sure to meet with your instructor to ensure that all the information necessary to complete your checkride is in the system. Also be sure to have your FAA Tracking Number (FTN). This is your unique identification number in the system. I found that I worried more about the IACRA part of what needed to get done leading up to the checkride than most everything else. It wasn't information I could put in a folder and the website is not the most user friendly; I had worries about something being amiss. Be sure to ask your instructor to help you ensure that all is in proper order.

In addition to having your information in the IACRA database correct, you'll also need the following for your checkride (always be sure to double-check though as requirements might change):

1. Knowledge test results—you should have your original printout to provide to the examiner
2. Photo ID—driver's license or passport
3. Student pilot certificate
4. Medical certificate
5. Logbook
6. Aircraft information
7. View limiting device (e.g., foggles)
8. *FAR/AIM* Book
9. Money for examiner fee

As with the IACRA information, be sure to spend time with your instructor reviewing your logbook one last time too. In addition to making sure all your time requirements have been met, you'll also want to make certain all endorsements are properly documented in its pages. If you are renting your checkride plane, you should review the maintenance logs several days in advance. It would stink to be ready to go and then find out the aircraft you intend to use is behind in required maintenance and inspections. Also ensure you have the weight and balance information for the aircraft you'll be flying. You'll need to show your ability to calculate the center of gravity for your flight and ensure that it is within limits. Every aircraft is different, so don't assume the information in a generic online POH is accurate; it almost certainly isn't.

Access to the *FAR/AIM* during the oral exam is crucial. While the exam is "open book" in the traditional sense, you will be expected to know a lot off the top of your head. It wouldn't be appropriate, for example, to look up whether you, as a non-instrumented rated private pilot, are allowed to fly in Class A airspace. If you need to reference chapter 3, section 2, page 2, paragraph (b) in the *AIM* to answer this question, that won't be a good look. But I was asked what the components of a flight review were and how long each of the components was. While I had memorized the frequency of flight reviews, I couldn't remember their structure and referenced the *FAR* to get the answer (a minimum of one hour of ground training and one hour of flight training if you're wondering). You will probably hear others seeking their PPL advocate vociferously for "tabbing" your *FAR/AIM* book so that you can quickly identify sections and, presumably, quickly find answers. I did this. But frankly, I didn't find it all that useful. The book has an index that takes you more quickly to a specific page and (hopefully) your answer. I wouldn't say that tabbing the book was wasted time; it did help me become more familiar with its structure and language. And often the exact thing you're looking for may not be in the index, so you'll need to have a sense of what it's related to. But don't forget about the index in the back. It's—in this aviator's opinion—a much better way to find what you need.

The oral portion

After I settled into a chair beside him in his small office, the DPE hummed to himself as he slowly went through my logbook and other documentation. I kept waiting for the humming to stop and him to say something like, "Uh-oh, looks like we have an issue...." Before he began his review of my logbook, he handed me the maintenance records for the airplane we'd be flying, N276ND and said, "Why don't you look through those while I review your paperwork." [78] Initially, I found it hard to focus on the maintenance records given my worry about his very methodical review of all I had brought to the exam.

Pretty quickly though, the concern about the DPE's review of my documentation gave way to a different anxiety. I began to go through the maintenance logs and realized I had *no idea* what I was looking at. I had never reviewed maintenance logs before. They were kept in the office at the flight school, but I never sat down with anybody to walk through an example of such records. It had never crossed my mind. This was something that I *hadn't* seen addressed in any of the hundreds of articles I'd read and basket of YouTube videos I'd watched. But as I mentioned earlier, the FAA tells you what you'll be tested on. And sure enough, in the *ACS* "Practical Test Checklist (Applicant)" there is this:

- ☐ Aircraft Maintenance Records:
    - ☐ Logbook Record of Airworthiness Inspections and AD Compliance

---

[78] In looking through the publicly available information on this Piper Warrior—a PA-28-161—I found out it was a 2000 model and that it had previously been owned by the University of North Dakota. UND is one of the largest university-based aviation programs in the country. The school started the year of my birth, 1968, with two Cessna 150s. Needless to say, N276ND had probably taken hundreds, if not thousands, of student pilots on their checkrides.

I knew that maintenance at my school was stellar—the owner was a former Air Force aircraft maintenance officer, there were several mechanics on staff, and several instructors who had been at other schools had mentioned to me that the maintenance systems at East Coast Aero Club were impeccable—so I had no real concerns about the records (or the aircraft) being problematic. But I was worried about being asked a question about them I couldn't answer. And that was a distinct possibility because the school did progressive maintenance; the records didn't show a single, annual inspection as I expected when learning about the regulations governing periodic inspections. Rather, they showed regular inspections at every 25 hours, done four times throughout the year (i.e., over a 100-hour period, thus meeting the 100-hour inspection requirement for flight school aircraft). Fortunately, I had some experience with progressive inspections and maintenance, having run a manufacturing company that used this sort of approach for equipment upkeep. So, when the DPE asked me a few questions about the pages, I was able to bumble my way through what seemed satisfactory answers. Thank God.

What other curveballs were coming? I wondered.

***

The DPE did ultimately find a problem in my logbook. A signature was missing. But fortunately, it was one from Connor, not Jeff. The DPE said it was no problem; we'd get Connor to sign the book when we saw him later. After the maintenance records questions, he referenced my written test report and asked me questions related to the areas identified as weaknesses. Fortunately, because I had only missed two questions, there wasn't a lot for him to latch on to. I've long since forgotten those questions but whatever they were, I knew the answers. Then, as predicted in the gouge, he pulled up satellite imagery of Hanscom and asked me about runway and taxiway markings.

After this exercise, it was on to the part of the oral exam that was more open-ended. We spent a good deal of time on the sectional chart. My DPE pointed to various portions of the map and asked questions about symbology and airspaces. Then, on to questions about the weather, including review of weather maps, winds aloft charts, terminal area forecasts (TAFs), METARs, AIRMETs (Airman's Meteorological Information), and SIGMETs (Significant Meteorological Information).

Next, we transitioned to the cross-country planning I had done per his request earlier in the week. He asked questions about the route I chose, the altitude, anticipated fuel use, and contingencies I had considered. He'd throw in a scenario every now and then to get me thinking in more "real world" terms, involving our friend, Joey-Bag-of-Donuts, regularly. I found his questions very thorough but fair. I appreciated greatly that he focused on information that would most matter to me as a "day-to-day" pilot. There were no queries on esoteric points that were so obscure I'd never run across them again in my time as a pilot. All told, from the time we started review of my paperwork until he ran out of questions for me, it was about two hours. He noted after I had answered his last question that I'd done very well in that portion of the practical test. That gave me a nice little confidence boost even as I was already feeling a little fatigued.

One piece of advice you've probably heard if you've been preparing for your checkride is to only answer the question that's asked by the DPE. I think this is very good advice. And for some of us—me included—it can be hard advice to follow. I'm one of those people who wants to fill awkward silences with the sound of my own voice. That may be a great at a cocktail party, but it's not so good when you are looking to make the oral portion of the exam as stress-free as possible. Expanding your commentary beyond the minimum required to answer a question from the DPE is opening a trap door to rabbit holes you probably don't want to explore. For example, if your examiner points to a "prog" chart (i.e., a prognostic weather chart) and asks what the blue line with blue triangles is, just say, "A cold front." Let them ask any follow-up questions. If you instead say, "A cold front, which is ultimately the result of a combination of the uneven heating of the Earth's surface and the Coriolis effect," then the DPE may decide that they'd like to know more about how, exactly, the two things combine to create a front since you suddenly look like a meteorologist. While I'd like to think you are prepared to venture into detailed territory after reading this book, it's a risky business. Just answer the question that's first asked. Do no more than that. Trying to demonstrate how smart you are will almost always lead to the opposite outcome.

Should we fly?

During the oral exam, I was glancing out the window in the examiner's office. It looked out over the tarmac. The winds were still strong, and daylight was becoming scarcer. It was about 12:30 local when we wrapped up the oral portion. Sunset would be at 16:19 local. I checked the latest METAR for KBED and it still showed winds in the 12- to 18-knot range. And those winds were out of the south. But for reasons I couldn't understand (and still don't), Runway 29 was still being used. The wind would be an almost direct crosswind.

The DPE had seen me reviewing the METAR and asked me how it looked. I said other than the wind, good. There were a few clouds, but they were high and sparse enough not to be a problem.

"Ready to go fly?" he asked.

I couldn't tell if this was a rhetorical question or a test of my aeronautical decision making. Every piece of content I had read or watched about taking a checkride emphasized that *you*, the examinee, are in charge. *You* are the pilot in command (PIC). *You* are the one who makes the calls on what to do when it comes to matters of safely flying your first true passenger—the DPE.

"Yeah, I think so," I said, not allowing myself to ruminate much longer lest I appear indecisive. Given all my training in windy conditions, I felt confident that I could execute the required maneuvers—particularly landings—to a satisfactory level. But I'd be lying if I didn't say part of my motivation to go was that I knew if we didn't, I would have to wait an unknown period to get back on the DPE's schedule. Part of me thought that if I said I wanted a discontinuance (meaning I wanted to stop the test and reengage later), then I'd be no worse off than if I failed a portion of the exam and had to retest. In other words, I'd have to reschedule either way. I didn't want to do that. This was admittedly not a healthy mentality. After the fact, the irony of it all wasn't lost on me: I had a form of "get-there-itis" when it came to my own checkride. While the DPE and I weren't going anywhere, the pressure was the same. I wanted to complete the mission that had been planned no matter what.

My DPE didn't seem to sense my inner turmoil.

"All right then, why don't you go start preflighting the airplane and I'll meet you out there."

It occurred to me later that the DPE probably expected that I'd be willing and able to fly in the breezy conditions. Given how much the air blew in our part of the world, you wouldn't be able to remain proficient if you stayed on the ground whenever there was a crosswind. That said, the fact that I'd be taking off and landing at KBED played a part in my call to continue. Landing on 29 at Hanscom, I'd have 7,000 feet of pavement that was 150 feet wide. The runway was surrounded by expansive, grassy areas. Deciding to land there was a very different call than deciding to land at Minute Man Airfield just 10 miles to the west. The only paved runway there was just over 3,000 feet long and only 48 feet wide. More critically, it was hemmed in by trees on all sides. This sort of analysis still plays a part in my ADM today.

The flying part of the practical exam

I got to the plane and took a deep breath. While the wind was noticeable, I was grateful for the warmth. It stinks to do a preflight in cold weather.

I had the checklist in hand and worked my way through it, forcing myself into the discipline of checking each item as I followed down the list with my thumb despite knowing every step by heart by this point. The DPE made it out to the airplane just as I was opening the cowling door to check the engine. He stood next to me and pointed to components in the cavity that contained the engine. I rattled off answers with confidence and conciseness. Connor had done a good job of preparing me for this part of the exam. Despite the DPE's seniority and vast experience relative to Connor, I found him less intimidating than my 21-year-old instructor.

The only uncomfortable moment during the preflight was when I ran my fingers along the leading edges of the prop while checking it.

"Why would you do *that*?" he asked with undisguised disappointment.

I looked at him, confused.

"I'm checking the prop for any cracks or large divots that might make it unsafe."

"You can do that by looking at the prop. *Never ever* touch it."

Suddenly, he did seem intimidating.

We were in the cockpit within a half-hour. I got situated and continued to run through the checklist. It was time to check the ATIS. Doing so ratcheted up my nerves. The focus on my preflight made me completely unaware of the runway being used. The ATIS reminded me: 29. *Why*? The wind was strong and almost right down Runway 23. But 23 was closed. All traffic was being routed to 29. I wasn't too nervous about a crosswind takeoff. But I was already thinking ahead to our return from maneuvers. I prayed that by the time we were back to Hanscom Runway 23 would be in use, otherwise I was looking at having to nail my landings with a wicked crosswind.

I went through a safety briefing that I had printed and laminated to the back of my checklist, noting rally points and procedures in the event of an emergency like an engine fire at startup.

I called ground to get taxi instructions. Sure enough, they were sending us to 29.

I had been warned by my CFIs and by others who had already taken the checkride to be sure to maintain a sterile cockpit once the airplane was moving—no talking about anything other than those things related to the execution and safety of the flight. I had drilled this into myself and even briefed it to the examiner during the safety briefing. While fatalities are few in taxi accidents, it's not unheard of, and damage to aircraft or vehicles is more common than you'd think. Distractions that lead to wrong turns or missed hold points are also a worry. I had steeled myself to rebuff any attempts by the DPE to engage me in friendly conversation, something I had heard he might do.

It was about 1.2 miles from the parking on the west ramp to the "hold short" line of Runway 29. The FAA recommends you taxi at no more than the equivalent of a "brisk walk." This is lunacy. A brisk walk for a fit, relatively tall person is about 4 mph. This is incredibly slow. I know of nobody who does this. Even those who think they do are underestimating their actual speed. An AOPA article by Dave Wilkerson on the topic of taxiing had a much more realistic and useful definition of an appropriate taxi speed: the speed at which an applicant can stop promptly while closing the throttle. I think that speed is likely somewhere around 10 mph, although of course it depends on the circumstances. Maneuvering in a tight area with a lot of people around, this may be way too fast. But on a straight taxiway with no traffic or adverse conditions (a wet surface, for example), it can be safe to go faster.

If I did taxi at 4 mph, the trip from my parking space to the end of Runway 29 would take around 20 minutes when accounting for slowing to take turns. I suspect I didn't go this slowly, probably somewhere between 4 mph and 10 mph. Whatever the speed, I would have preferred that the taxi was shorter because there was a *lot* of dead time in the cockpit. And as we know, I'm a social guy. I had to be on my guard against getting drawn into pleasant chit-chat.

"What do you do for a living?" the DPE asked.

And, like a golden retriever who told himself over and over again that he will not chase the ball no matter how enticing it might seem, but who just can't help himself ...

"I work for a university ..."

Seriously? Am I an idiot?

I caught myself.

"But let's talk about that when we don't need to maintain a sterile cockpit anymore," I added as if I had planned all along to give him a teaser about my career only to leave him wondering.

He said nothing.

One thing that will help you not fall into a similar trap during your own checkride is to verbalize *everything*; then there is no room for conversation anyway. The DPE isn't a mind reader. Saying what you are doing and thinking out loud lets them know that you are doing what you're supposed to do.

"Looking left, looking right, all clear," I said as we came upon an intersection with another taxiway.

"This is a hotspot," I added as I made the turn, "and we want to continue east on Taxiway Echo."

During the taxi, I made it a point to position my flight controls with the wind direction in mind. While always a good practice, it's especially important on windy days. Having the flight controls in an incorrect position could, in strong gusts, allow the lifting of a wing, moving the plane in a direction you didn't intend. There are a lot of diagrams that explain which aileron should be up and which should be down, in combination with an elevator position. But the easiest way to remember what to do is summed up in this way: *dive away from a wind from behind, turn into a wind from the front.* Diving away from a wind that is coming at you from the three o'clock to the nine o'clock position ensures that the elevator is down, preventing wind from easily getting under your horizontal stabilizer and lifting the tail. It also puts the aileron on the side of the wind in the "down" position; this serves the same purpose in helping the wing stay down as the wind is forced up and over the control surface. When wind is coming at you from the nine o'clock position to the three o'clock position, just turn into it with the elevator in a neutral position. This puts the aileron on the side of the wind in an up position, helping force that wing down.

After what seemed an interminable commute, we finally arrived at the runup area. I turned the aircraft into the wind and stopped it to begin my pre-takeoff and runup checklist.

V_x and V_y

I continued to verbalize everything, demonstrating that I was following the checklist precisely. My DPE had asked that we start with a short-field takeoff with an imaginary obstacle at the end of the runway. I briefed to him on what we would do to accomplish this maneuver: "flaps will be at 25°; rotate at 52 knots; climb out at V_x or 57 knots; when clear of obstacle, lower nose and accelerate to V_y; retract flaps slowly." A short-field takeoff is exactly what it sounds like—a procedure to quickly get off the ground if the amount of runway in front of you is limited and an obstacle (typically defined in performance charts as a 50-foot-tall object) at the end of the runway means you must gain altitude in the shortest distance possible. V_x is the airspeed to pitch for that will give you the maximum altitude gained over the shortest *distance* along the ground, otherwise known as best angle of climb. You've undoubtedly also heard of V_y. This is the speed to pitch for when trying to get the most altitude in the shortest amount of *time,* also known as best rate of climb. There is often confusion around this dynamic duo for two reasons. First, people simply forget which is which. To help with that, here's another original ditty to remember:

For *quickest time* to the sky,
Pitch your plane to go V_y,

Runway short, 50-foot objects?
Pitch your plane to go V_x.

The other confusion centers around what the difference is between them. Why isn't there just one speed? Getting into the math of it is too much for this book. That would require discussions of excess thrust, an airplane's power curve, and several more esoteric things. But to oversimplify, your airspeed when climbing out at full power is a proxy for your angle of climb, which defines how fast you'll get to a given altitude or how much distance you'll travel over the ground to that altitude.

A Piper Warrior III has a V_x of 57 knots with 25° of flaps. In other words, 57 knots at that flap setting will give you the optimal climb performance at full power relative to distance traveled over the ground. Pitch any higher and your speed over the ground will decrease as more of your velocity translates to the "up" vector vs. the "forward" vector. This should be a good thing. After all, you're trying to get the most altitude in the shortest distance over the ground. But the problem is pitching higher increases your angle of attack, bringing you closer to a stall, degrading climb performance. If you pitch your nose down from V_x (i.e., increase your airspeed), you will increase your forward speed at the expense of vertical speed. This means at a given horizontal distance, you'll be at a lower altitude than if you had just stuck with V_x.

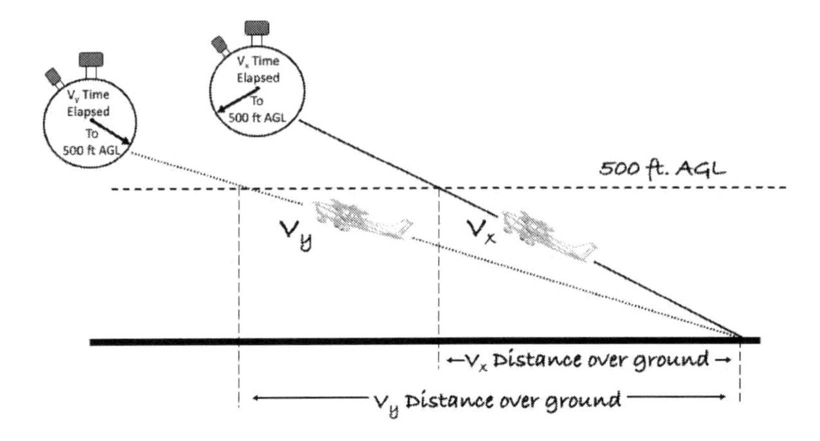

The logic to understand V_y is the same. With no flaps, V_y is 79 knots in the Warrior III. This gives you the maximum altitude gain in a unit of time. You have an instrument in the cockpit that can show you this; the vertical speed indicator shows your rate of climb in feet per minute. If you want to get as high as possible as fast as possible, you'll want to see the needle on that instrument deflect as far "up" as possible (but not so far up that it's not sustainable and you put yourself on the verge of a stall). V_y does that for you.

I was cleared onto Runway 29 and looked down final to be sure nobody was coming. I pulled onto the centerline and verified that the heading on the compass matched the runway heading. I asked the DPE if he was ready to go, and he said yes. I reminded myself—out loud again—to get my heels on the floor to ensure I wasn't going to be inadvertently riding the brakes as we accelerated. I ran up the throttle, and we were on our way.

The takeoff went well, and as we transitioned from V_x to V_y the tower asked me to turn on course to clear the departure leg of the runway for a higher-speed aircraft taking the runway behind me. We were high enough to make the turn safely, so I looked to the right to clear my turn and then turned the yoke as I lightly pushed the right rudder. As I began the course change, I parroted back to the tower, "Turn right, proceed on course, Warrior 276 November Delta."

At that point, my examiner asked me what altitude I had chosen to cruise at in my cross-country planning. But his voice cutout on the intercom and I didn't hear him.

"Could you repeat that please?" I asked.

"What cruise altitude did you choose for your cross-country flight?" he said again.

"Eight thousand, five hundred," I responded. I had chosen a higher altitude than I otherwise would have because the cross-country route would take me out over the mountains of Vermont and eastern New York. I wanted plenty of clearance and, given how wooded and mountainous the terrain was, time to make some decisions if I lost my engine.

"Okay, establish your climb to that altitude and make your way to your first checkpoint."

"Okay, will do," I responded.

This exchange was innocent enough. In fact, I was feeling *really* comfortable with how things were going. But the whole while, something was nagging me. *But what?*

And then it hit me.

Crap!

I had never let up on the mic button after acknowledging the call to the tower to proceed on course. I was pressing so hard on that button I'm surprised I didn't pulverize the yoke into a cloud of whatever-they-make-yokes-out-of dust.

As soon as I released the button, I heard the tower controller in my ears.

"There's an aircraft with an open mic on the frequency …"

She must have realized that the frequency was now free and immediately started giving directions to aircraft. Directions, no doubt, that had been delayed in delivery by some idiot's inability to release the mic button after he was done talking.

But here's the funny thing: the DPE didn't know I was the one with the open mic. How could he? As far as he knew, there was just a lull on 118.5 Mhz. Of all the mistakes I could have made, this was a "good" one. I pretended like nothing had happened and made my way to my first checkpoint. But it was clearly an indication of my nerves, no matter how I tried to convince myself they weren't a problem.

After flying northwest of the field about 15 miles, the DPE told me to put on my foggles and navigate to the diversion he had me plan for during the oral portion of the exam. The diversion wasn't an actual airport, but there was a defining landmark I was to instead use as my destination. As is common, the wind at altitude was even stronger than at Hanscom; it was 45 knots at 3,000 feet MSL. But I had calculated my wind correction angle earlier on the ground and turned N276ND to the heading I had written down in my navigation log. I also started the timer on my watch. The distance to the diversion location was short, on the order of 12 nm, so the time to get to it would also be short. When I earlier calculated the time to the diversion, I had reduced it from what the math told me. This was because of a great piece of advice another aviator had given me. If I was early to the diversion waypoint, it might be hard to find the visual marker the DPE had asked me to look for since it would be behind or under the airplane. Better to take the foggles off 30 seconds before my calculations said I should to ensure the landmark would be in front of me.

The stopwatch I had started reached the time I had decided would put me just short of the waypoint.

"I think we're there," I said to the DPE.

"Okay, take off your foggles and identify the diversion point."

I did as he instructed and slipped the foggles into the small pocket near the floor on the left side of the cockpit. I began scanning the terrain below.

I couldn't see what I was looking for.

"Do you see it?" the DPE asked.

"I don't," I said, feeling sweat start to squeeze from my skin.

"Isn't that it right there?" the DPE said, rising in his seat, craning his neck, and pointing across my body to our eleven o'clock position.

Yep, that was it. At first, I was embarrassed. How in the world had I missed it? My thumb on the mic button had a mind of its own and now my eyes weren't working? I didn't deserve to pass this exam.

"Good job," he said, apparently unaware of my own self-loathing. But then I realized—I had nailed it. On a very windy day, I had calculated the correct heading at a given altitude and the precise time to put me over a location I had identified on a map. The loathing quickly evaporated, and pride supplanted it. I was feeling good.

Maneuvers time

"Okay, let's fly a heading of about 275° and head out to the practice area," the DPE said, seemingly satisfied I had successfully completed the navigation to the diversion.

We didn't fly to the practice area in normal cruise. Instead, he asked me to configure the aircraft for slow flight in a landing configuration, meaning flaps would be deployed as if landing the aircraft. Slow flight is required in training and the exam because of the importance of understanding how it "feels." That feeling is often called "mushing." It's an appropriate word. When at slow speeds and high angles of attack—as when you are landing and taking off—the control surfaces aren't as effective; you must deflect the yoke or rudder pedals relatively far for little roll, pitch, or yaw. The controls feel "mushy." It's easy to visualize why. We just need to once again reference Newton's third law. If the air is flowing more slowly over the aircraft and therefore hitting the control surfaces with less force, there's less of an "equal and opposite" reaction imparted to the wing (in the case of the ailerons) or tail (in the case of the rudder or elevator). Even large control inputs do not have meaningful effect. This feeling is one to know well since takeoffs and landings are when you're likely to feel sluggishness in the controls and you'll be low to the ground. That said, the slow flight maneuver you'll be asked to perform is done for a prolonged period on the edge of a stall, and you will often be asked to make turns while in that state. It can feel uncomfortable, which is sort of the point. Doing slow flight should help you get a sense of the "feel" of flying dangerously slow. It is also why you should do the maneuver at altitude; if you'd don't manage your angle of attack or yaw well, you'll want time to recover.

I executed slow flight in both the landing and takeoff configurations well. By the time I was done with both, we were over the practice area and began the rest of our maneuvers. Those included stall recovery, steep turns, turns around a point, and an emergency descent and simulated off-airport landing. These made me nervous, even more so than the landings to come back at Hanscom. I suspect there were two reasons for this. First,

I had more practice at landings than maneuvers. I *had* to land every time I trained. I didn't have to do steep turns. The other reason was that a runway is a clear and distinct aim point and it's at 0 feet AGL, all the time. While you may move in three dimensions to get to a landing spot, success was more or less measured in two dimensions: how far down the runway you touched down and how close you were to the centerline.[79] Not so with maneuvers done at altitude.

For example, the *ACS* says that for a steep turn to be executed successfully, the pilot must hold a coordinated turn with a bank of 45° ± 5°, hold speed to within ± 5 knots throughout the turn, hold altitude to within ± 100 feet, and roll out of the turn no more than 5° from the starting heading. That's a *lot* to manage. This is especially true because any input in one dimension almost always has an impact on the others.

Once at the practice area, I did my clearing turns to ensure no other traffic was around. I verbalized even here: "Clearing turn to the right … no traffic to the right…." But no matter what I did that made me think I was executing clearing turns well and keeping my head out the cockpit through all of this, the DPE let me know it wasn't enough. This drove me crazy. I had convinced myself that I wouldn't do what all those other third-rate test-takers had done to earn the ire of the DPE during their checkrides. I would do clearing turns for my clearing turns! I would have my eyes outside the airplane so much the only way they could have been outside more is if each of them was

---

[79] There are, of course, other requirements related to minimal float, appropriate speed, etc. But after so many landings, you'll probably find that virtually all your mental energy is focused on hitting an aim point and keeping the plane down the centerline.

surgically attached to the cowling. But no amount of effort in this regard seemed to satisfy him. This was persistent feedback throughout the checkride, and its persistence led to persistent anxiety about my efforts. Was I going to fail this thing?

When my stall maneuvers were done, the DPE noted that we were no longer over our emergency landing spot, a clear field not too far from a reservoir. After he pointed this out to me, he told me to make my way to the airspace over that field. I turned the plane to the left.

"I have the controls," the DPE said suddenly.

I reacted instantly, as I'd been trained, released the controls, and parroted back, "You have the controls."

Completing the ritual of positive transfer of controls, the DPE parroted back my parroting: "I have the controls."

He then swung the Warrior into a much steeper turn, accelerating our return to the practice area.

"Let's get back more quickly," he said. It was becoming clear to me that he was worried about the dissipating daylight and wanted to be sure we minimized time between maneuvers, giving us the best shot at completing our activities while the sun was still up.[80] While this all made sense, I couldn't help but let his impatience with my turn lead me to believe once again that I was making a mess of the checkride.

Once back over our emergency landing spot, it was time for steep turns. The anxiety I described about doing these turns a couple of paragraphs ago was real, but it was significantly less than it was earlier in my training. Like so much of flying, you

---

[80] This made me curious: *Can* you do a checkride at night? In a 2018 AOPA article by designated pilot examiner Bob Schmelzer, he notes that it is indeed permissible to conduct a checkride at night. Who knew?

increase your chances of success by finding a configuration for the airplane and visual reference points that take some of the variability out of the maneuver. When it came to steep turns, the configuration item that mattered most was engine power. Know what your tachometer needs to read to ensure you will maintain altitude when you are banked at a particular angle and pulling through the turn. Next, have a reference point on the cowling or cockpit glare shield so that when that point and the horizon meet, you know that the nose is pointed where it needs to be to keep you at a consistent altitude during the turn. I had been trained to do this and while it took a little while to master, it worked. I felt good as I rolled out of the turn on the exact same heading I had started it on. I was equally flawless in a turn the other direction. Of all the maneuvers I did that day, I'm most proud of my steep turns.

Turns around a point were next. This, on the surface, is a simple maneuver. Even the description in the *ACS* of what's required to execute the maneuver is tame compared to that for steep turns: *enter at an appropriate distance from the reference point, 600 to 1,000 feet AGL and maintain altitude within ± 100 feet and airspeed within ± 10 knots.* The issue is that in the "Knowledge" portion of the *ACS* dealing with ground reference maneuvers, it is noted that "the applicant should demonstrate knowledge of the effect of wind on ground track and relation to a ground reference point." Did I mention it was very windy that day?

We entered the maneuver at 1,000 feet AGL on the downwind. I knew the wind at the surface was somewhere between 15 and 25 knots. And I knew that at 3,000 feet it was about 45 knots. It was reasonable to assume then that I was dealing with something on the order of a 30-knot wind at 1,000 feet AGL (or about 1,500 feet MSL near the practice area). So, the challenge was going to be flying a ground track that was circular and maintained the reference point at the center of the circle being drawn (i.e., making the reference point the "hub" of the circle). Windy conditions mean that if you hold a constant bank when doing the maneuver, the ground track will not be a circle. You therefore need to bank more when on the downwind portion of the turn and less when on the upwind. And because turning in a circle means that the angle at which the wind is hitting the airplane changes with every degree of turn, your control inputs must be *constantly* changing. This makes turns around a point in windy conditions much more difficult than steep turns. In a steep turn, it doesn't matter what the wind is doing because you don't care about your ground track. You are like a fishing trawler motoring in a circle on the open ocean, oblivious to the current moving you in one direction or the other. That's one reason you can predetermine power settings and visual references on the airplane in preparation for a steep turn. You can't do the same when it comes to turns around a point (at least with the same precision) because the wind *does* matter. It matters a lot. For this reason, it's perhaps one of the most "fly-by-the-seat-of-your-pants" maneuvers.

We used a small building in an open field as our "hub." While I did look at the building periodically, I learned that fixating on it made it harder to fly a consistent, circular ground track. Instead, I employed a trick that Connor had taught me. I looked ahead of me and envisioned the arc I wanted to fly, picked a landmark that fell on that imaginary arc, and then I aimed for *that.* As I got close to the chosen landmark, I'd look ahead and pick another. This took a lot of the thinking out of when to bank more and when to bank less based on where I thought the wind was coming from at any given moment. Instead of doing that mental exercise, I just aimed for something fixed on the ground and made the plane go there. This was highly effective. The DPE seemed pleased.

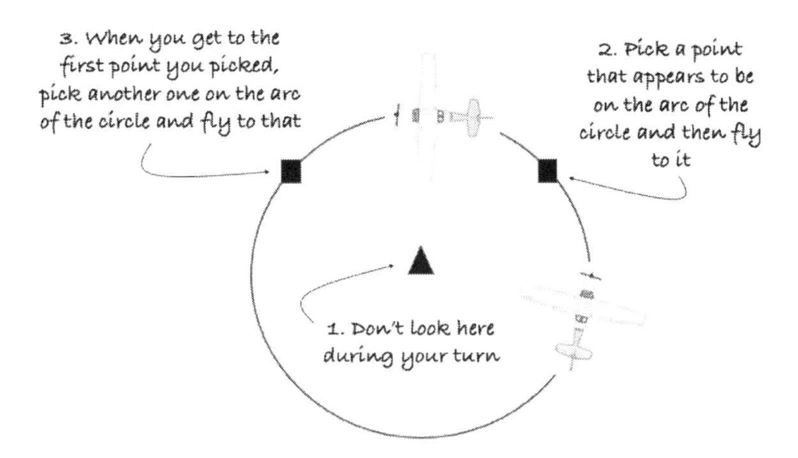

We climbed now so that we could practice an emergency descent and simulated off-airport landing. I've noticed that emergency descents are not always covered in a POH. To the extent they do exist, they are usually buried in the procedure for an engine fire in flight. This makes sense given the gravity of such an emergency (you want to get down ASAP) and because a rapid descent that forces significant amounts of air to flow over the engine might blow out the fire. A POH for a 1972 Cessna 172S references an emergency descent in the engine fire procedures section by saying that the pilot should establish and airspeed of 100 KIAS and, "if the fire is not extinguished, increase glide speed to find an airspeed, within airspeed limitations, which will provide for an incombustible mixture." You'll note that you are gliding because when an engine fire occurs you want to remove all sources of combustion (sparks from electrical systems and the spark plugs) and fuel (this is why you'll turn off the fuel pump and engage the fuel shutoff valve). The Warrior III POH—for models from 1995 and up—doesn't even provide the information the Cessna POH does. In fact, it seems misleading.

If you followed the instructions in the emergency section of the POH as written, you'll do all the things to cutoff sources of sparks and fuel as we just discussed, but then proceed with the "power off landing" procedure. But this procedure tells you to establish a best glide and look for an emergency landing spot. Establishing best glide is the right thing to do if you're not on fire as it gives you the most time in the air to find an acceptable landing spot. But when your plane is on fire, you aren't much interested in taking your time to get to the ground.

At 3,000 feet MSL, the DPE pulled back the throttle and said, "Your engine is on fire."

The image of the cockpit I had printed and drawn arrows on denoting the order of events for this emergency popped into my mind's eye. My left hand reached to the floor to simulate turning the fuel selector to "off." Then I simulated pulling back the throttle and the mixture, turning off the fuel pump, and ensuring the heater and defroster were turned off (they pull heated air from around the engine, so they provide a nice conduit for smoke to get from the outside to the inside of the cabin). I also simulated turning off the master and alt switches. Then, I ran through my "ABCDE" acronym.

A—establish the appropriate **airspeed**. This was around 120 knots in a descending turn.

B—find **best** place to land. As soon as I was done with the in-cabin procedures, my eyes went to the outside to find a suitable off-airport landing site. I identified a field that I was going to make a play for.

C—reference the **checklist**. I quickly grabbed my checklist and verified I had done all I needed to.

D—**declare** an emergency. This step was unnecessary in this scenario. I had simulated turning off the master and alt switches, which meant there would be no power to the radios.

E—prepare to **exit**. Some aircraft checklists will instruct you to open the door, helping ensure that if metal gets twisted you don't end up trapped in the airplane.

I stepped through my ABCDE acronym while putting the plane into a dive that brought the airspeed up to about 120 knots. I corkscrewed down while I did this. It's not always necessary to descend with a turn. I did so because the terrain around the emergency landing spot I had identified was heavily wooded for miles. I didn't want to fly away from that location during my (simulated) emergency, so turning over the field ensured that when—hopefully—the fire was out and I had given up a lot of altitude, the field was still reachable.

"You gonna make the field?" the DPE asked.

"Yes," I said with confidence. He seemed to agree and told me to recover from the maneuver.

"Let's head back to Hanscom," he said. I turned the plane back to the northeast.

Checkride landings

I was feeling relatively good as we cruised back to Hanscom. The maneuvers seemed to go well. The DPE never said anything that indicated I had failed one of the required elements of the practical test. The only thing that was bothering me was his continual commentary regarding clearing turns and looking out the window for traffic. He was a see and avoid pit bull. In retrospect, I think he might have intentionally over-indexed when it came to commentary on this aspect of flying because he knew how easy it was to slip into complacency when it came to looking for collision threats. With all the technology available to us in the cockpit, it's easy to get mesmerized by it, locking on to it with our eyes like a dog eying a piece of bacon. As I discussed earlier, I'm a huge fan of ADS-B connected to in-cockpit displays denoting the location of traffic. Having access to this technology is hugely helpful when it comes to situational awareness. But in the end, you'll still need to look out the window and identify other aircraft the "old fashioned way"—with your eyeballs—if you are to ensure you don't collide with them. Also, visual contact with an aircraft is the only kind of contact a controller who warns you about traffic cares about. Telling them that you see the target on your ADS-B display is not sufficient.

After demonstrating that I knew how to tune in the Boston VOR and interpret what the needle on the CDI (course deviation indicator) was telling me, I tuned the second comm radio to the ATIS. Ugh. Winds were still very high, and Runway 23 was still not available for use. All traffic was landing Runway 29. The winds were reported at 235 at 18 knots gusting to 25 knots. I thought I was going to barf.

"November 267 November Delta, enter a right downwind for Runway 29." Damn. Not only would I have a wicked crosswind, but I'd be in a right traffic pattern. While I'd flown both right- and left-handed patterns, I had significantly more practice with left-hand patterns.

The sun was getting low.

I flew on a northeasterly heading then, once well north of the extended centerline of Runway 29, headed east until I turned slightly right for a 45° entry to the right downwind for 29. At midfield, the DPE pressed the comm button on the yoke in front of him and said, "Wind check?"

The tower responded: "Winds 230 at 20, gusting 25."

The DPE started punching at numbers on his phone, doing a calculation.

"You're gonna have a 17-knot crosswind," he said. It wasn't lost on me that the Warrior's maximum demonstrated crosswind was 17 knots. It didn't mean the plane couldn't be landed in a stronger wind; the POH notes emphatically that "demonstrated crosswind values are NOT limitations." But that was little comfort. I once again had a nagging worry creep into me that he was testing my aeronautical decision making. I imagined he might have been thinking, *is this idiot really going to continue this practical exam with crosswinds like this?* It wasn't an unreasonable assumption. I still had a chance to ask for a discontinuance.

One of the possible outcomes of a practical exam, a discontinuance is not a failure. It's essentially a "press pause" option and is available for three reasons: illness of the examinee or the examiner, mechanical issues with the airplane, or weather. While either you or the DPE can press the pause button due to illness, the examiner will likely rely on you to make the stand-down call if mechanical issues or the weather are a concern. Don't be afraid to use this option if you feel things are getting out of your comfort zone. You won't be required to redo maneuvers you successfully completed up until the point you ask for a discontinuance. Admittedly, rescheduling can be a pain. DPEs are typically in high demand, and trying to find that magical time when weather along with DPE and aircraft availability line up may make you think twice about playing the discontinuance card. But it's better than pressing your luck and failing the checkride or, worse, putting yourself, the plane, and the DPE in jeopardy.

Speaking of failing a checkride, if you do fail a maneuver, the examiner will let you know immediately. You won't go through the entire test only to be surprised after you shut down the engine on the ramp. That said, you will be given the option to continue the exam once you are notified of a failure unless the DPE thinks continuing would compromise safety. Anything you complete satisfactorily from that point forward still counts; you won't be required to repeat everything the next time out. Instead, you'll have to get additional training on the failed portions and then repeat just those items the next time you are able to get with the DPE.[81] Generally speaking, you should always continue the exam even if notified of a failure in the middle of it. Every item you successfully execute that day is one less item you'll have to do again when you reschedule with the DPE. And since, after an initial failure, you know you'll have to get on the DPE's calendar again, any subsequent failure can be retested during that session (although we obviously hope there won't be additional missteps). So, you might as well try and get through as much of it as possible.

When the DPE notified me of the crosswind I'd be facing, I never considered taking the discontinuance option. As I've mentioned before, my comfort with crosswind landings was high. I tried to psych myself into believing that this was, as we used to say in the military, an opportunity to excel. I would show the DPE how adept I was at landing in less-than-ideal conditions.

While approaching Hanscom, he told me to start with a regular landing and then we'd stay in the traffic pattern and finish up the exam with a soft-field landing and takeoff and end with a short-field landing.

---

[81] To take advantage of the opportunity to just repeat the items you failed, you must fly with a DPE again within 60 days of your failure or you *will* have to complete the entire practical exam.

I verbalized my BC-GUMPS check, ensuring my fuel pump was on, carb heat was off (unlike other aircraft, carb heat was not required during landing in the Warrior), the fuel selector was on a tank with sufficient gas, mixture was rich, power was where I wanted it, and our seatbelts were secure.

"Abeam the numbers, airspeed in the white arc, first notch of flaps," I said as the threshold of 29 passed off our right.

Fortunately, I was able to fly a standard pattern; traffic was light and therefore the need to extend my downwind or perform some other aerial gymnastics wasn't necessary. I deployed another notch of flaps when on base. I was feeling good about my speed and descent rate as I turned final. The runway lights were on at the field. The sun had dipped below the horizon leaving a bright glow that backlit the hills to the west.

I flew my normal approach by aiming the nose of the airplane 400 feet short of where I planned to touchdown. I would keep that point fixed in my windshield until I pulled back on the yoke to begin my roundout. I knew that point was exactly 400 feet in front of where I hoped to touch down because the aimpoint was two runway stripes in front of my touchdown point. Each runway stripe is 120 feet long and the gap between them is 80 feet. Therefore, two stripes and two gaps constitute 400 feet of distance. My desired touchdown point was the stretch of pavement punctuated by what are colloquially known as the "thousand-footers." These markings are large rectangles painted white and straddling the centerline about 1,000 feet from the end of the runway, thus their name.[82] They are slightly longer than

---

[82] The FAA calls the thousand footers "aiming points." This seems an arbitrary designation as there's no requirement that pilots aim for them. They

the runway stripes at 150 feet and much wider at 20 feet. Their width and the fact that they straddle the centerline mean that they are easy to see in your peripheral vision even when the airplane is over them.

I performed a side slip to offset the effects of the crosswind, dipping my left wing while applying right rudder to keep the plane aligned on the centerline. I touched down right in the center of the thousand footers, pulled up the flaps and applied full throttle to climb back out. I was feeling good.

"Good," the DPE said. "Let's set up for a soft-field landing this time around with a stop-and-go, and on the go we'll do a soft-field takeoff."

We received permission for the stop-and-go and I executed the soft-field procedures well. One more thing left to go: the short-field landing.

Of all the landings, the short-field landing is the most challenging. To meet the standards, the examinee must land no more than 200 feet beyond a predetermined location on the runway and zero feet in front of it. This is to simulate a scenario where you might need your touchdown point to be very close to the runway threshold (or the beginning of a nice off-airport landing spot in the event of an emergency). Being short might mean touching down somewhere other than the runway surface. I told the DPE that my touchdown point would be the beginning of the thousand-footers. Since these were 150 feet long, that meant that I could land anywhere on them or up to 50 feet beyond the end of them. But I couldn't be short of them.

---

are special though in that they represent the area where the glide slope of an instrument landing approach (ILS) signal intersects the runway.

My approach felt good, but because of some gusts I was carrying a little extra speed. And wouldn't you know it? The gusts seemed to disappear as I began to flare. My ground speed was higher than I expected as the component of headwind I had counted on never materialized. I saw the thousand-footers slip under the plane, the ends of them visible just over the cowling, looming like cliffs to oblivion. I cut the power completely and relaxed my flare, hoping the airplane would sink and the wheels would get on the ground in an allowable range. We hit hard and flat, but within 50 feet of the end of the thousand-footers.

"What are you doing?!" the DPE exclaimed. He spoke again before I could answer, which was a good thing since I wasn't sure what to say.

"You've got to keep the nose up our you risk a prop strike!" he said, clearly agitated.

Well, I've done it, I thought. No way I'm passing now.

We taxied in silence back to the ramp. While a sterile cockpit during taxi is something a DPE will like to see, his agitation at my landing made the silence feel more like the weighty one that hangs between a couple in a car after a fight, not like a best practice for safe aviating.

When we got back to East Coast Aero Club's ramp, all the other planes from the school were parked, a day of flying done. On that section of the ramp designated for Warrior parking, there were twelve parking spots, six in one row and six in another (although, as will become obvious in a moment, I had never actually counted how many spots there were on each side). The rows of planes faced each other across a wide taxi area with room behind the parking spots allowing planes to taxi and pull through to park; no need to push planes back into their spots.

As we approached parking, it appeared there was a spot on the far end of one of the rows. We taxied behind the planes already tucked away for the night and planned to pull into the last slot located just before a rope cordoning off the ramp area beyond.

Uh oh.

My heart sank. We were wrong; there was no last parking spot. Just a dead end. We were stuck. Planes on our left, a blocked area in front of us, grass and taxi lighting to our right.

The DPE was gracious, saying repeatedly that he thought there was one spot left too. He told me to shut down the engine and then got out and pushed the plane backward while I pressed hard on the right rudder pedal, forcing the tail to swing out over the grass lining the side of the taxi area opposite the parked planes. We essentially had to do a 10-point turn with the engine off as the DPE ran from the front of the wing to the back of the wing and I alternated between right and left rudder. Eventually, he got back into the plane, breathing hard from having to push the aircraft around in sharp turns with my 200-pound frame sitting in it. I started the engine and taxied back to the beginning of the row where there was a spot we had inexplicably passed up earlier. Once parked, I completed the last few checklist items and shut down the engine.

"Well," the DPE said, "are you ready to take your wife on some trips with you?"

I think I might have said, "Yes sir!" as if responding to one of my old military commanders. But I'm not sure. I was too elated to note the specifics of my reaction.

He shook my hand in the cockpit and then said that we should get out while there was a little light left so he could get a picture of me in front of the plane. My watch is clearly visible in the photo; it was 4:52 pm local time. The sun had set 33 minutes earlier.

I was a pilot.

Epilogue

I finally did buy that plane. It wasn't a Cessna 172 but a Cessna 177—a Cardinal. I bought into a partnership that has only one other owner who also happens to be a great guy. I completed the transaction about a year and a half after I got my certificate. The plane is currently based at 6B6, Minuteman Air Field. It's a great little airport that reminds me of what all airports used to be: quiet, almost bucolic settings punctuated only by the sounds of birds and piston engines.

In the summer and fall, my wife and I love to hop in the plane and fly to Martha's Vineyard. Katama Airpark (1B2) is usually our destination. It's even more of a throwback than 6B6—only grass runways and taxiways at this airfield. It should be a bucket list destination for every pilot. There's a parking area that gets you to within 20 yards of the beach. Beautiful Edgardtown, the largest town on the island, is a short taxi ride away. It was here that much of *Jaws* was filmed. It's 77 miles from our house to Edgartown as the crow flies. On a good day, the trip by car and ferry would be three hours, and on a bad day it could be five. We make the flight from 6B6 in about 40 minutes. Often, Mary and

I will grab lunch in Edgartown and it's always a bit of a thrill to tell those we meet that we flew ourselves to the island and that we'd be back home by dinnertime.

Since getting my certificate, I've gotten to fly with my 77-year-old father on two very meaningful flights. One was a trip to Hudson Valley Regional Airport (KPOU) in Poughkeepsie, New York. While KPOU has a great restaurant, that was only a secondary purpose of the visit. We made our way there by flying up the Hudson River Valley, over West Point. My father graduated from the military academy in 1967 and began his life of aviation immediately after, never realizing that his passion for flight would end with several trips into space. Flying with him over the place where his aviation journey began was meaningful for both of us. If he hadn't honed his dream to fly while living on the banks of the Hudson, my own passion for flight would probably never have materialized.

More recently, we spent three days in the summer flying around Cape Cod and Martha's Vineyard. One especially fun memory of that trip was landing at Provincetown, Massachusetts (KPVC), on the very tip of Cape Cod. The views of the entire cape from the air on a clear day are spectacular and, if you look straight down at where the sand meets the water, you'd be forgiven if you thought you were flying over an island in the Caribbean (although a dip into that chilly water would remind you of where you were!). To call Provincetown eclectic is to undersell just how unique it is. We caught a ride into town and had lunch while we looked out over Cape Cod Bay and then walked the streets of the village where people-watching is an Olympic sport.

Most don't know that the location of modern-day Provincetown was where the Pilgrims first landed in the New

World. They spent about two months on the (then) desolate spit of land before making their way across the bay to Plymouth. Their landing on Cape Cod came 287 years before Lord Haldane questioned the success of the Wright Brothers. As doubtful as Lord Haldane was, imagine how skeptical those Pilgrims would have been if, in 1620, somebody told them that one day two men would fly a machine to a location near where they anchored and that one of them had flown into space?

Any significantly advance technology is indistinguishable from magic….

Dad flew back (via a commercial airliner) to his hometown of Albuquerque, New Mexico, a couple of days after our flying adventure. Shortly thereafter, I sat down to my email to catch up on a few things. Waiting for me was this note:

Pat,

Thank you again for the incredible weekend. By far, it was the most wonderful experience a father can share with a son. It was a REAL bucket list item…though one which is self-replicating. At each landing I was reminded of my shuttle days:

Before flight: 'God, if you get me into space this one time, I'll never ask for anything again!'

After flight: 'God, if you get me into space AGAIN, I'll never ask for anything more!'

Repeat.

It's going to be that way with flying with you…a need never satisfied. That weekend wasn't enough. I can't wait to come back and fly with you. And get mom up there to fly, too. A bluebird day in the fall with a trip over the foliage to New Hampshire for lunch. That's now on the bucket list.

And repeat!

Seriously, I am so happy you pursued your pilot's license and bought the plane. You can't know how deeply I was affected by our time in the cockpit. I had long resolved I would never again live the joy of flight from a cockpit position, but you resurrected it. Every moment of this weekend was special: pushing the plane from the hangar into the deserted stillness of the early morning; staring into the sky and sharing opinions with you on the weather (while correctly calling the altitude of the lowest clouds :)*); watching you run through the preflight (which reminded me of my F-4 days); hearing the cough of the engine; feeling the vibrations of startup; watching the blur of the prop; thrilling to the rush of the plane down the runway; experiencing that most unique of human moments, when you brought back the yoke and the altimeter spun the earth away; of being surprised I was still fluent in the ATC language of flight (I guess, once learned, it never leaves you); and, watching a perfect landing pattern conclude with that sweet song of touchdown…a stall warning followed by the chirping kiss of the tires on the runway….*

It was ALL transcendental. Thank you! Thank you! Thank you!

I love you! And let's go flying again soon!!!

Dad

That email alone made the journey worth it. I hope that each of you can find your own slice of joy in this passion we share.

Acknowledgements

So many helped get this book into your hands through their diligent counsel and keen eye for detail.

First, thanks to Jennifer Whitley, my editor, who offered incredibly valuable and prompt feedback. She made this book better. When I went looking for an editor, I never thought I'd find a literary mind with an aviation background. Jen was that rarest of unicorns. I am so grateful for her efforts to make these pages better than they otherwise would have been.

Writing a book about a technical topic is always fraught. There are so many out there who, as the saying goes, have forgotten more on a given topic than I've ever learned. The value of feedback from experienced pilots cannot be overstated. Jim Dubela, an Air Force veteran and captain for a major airline, Jim Henry, an accomplished CFI, John Zimmerman, an experienced GA pilot, and John Williams, all provided corrections and suggestions when they had so much else to do. John Zimmerman is the host of *Pilot's Discretion Podcast* from Sporty's. He's one of the best interviewers out there when it comes to engaging with members of our flying community. I highly recommend his program.

In addition to these fine pilots, my cousin, Vince Sei, a former Air Force officer and aviator, also provided keen insights. Someday I hope to fly with him in his Vision Jet (hint, hint, Vince!).

The lessons I've learned from my close circle of flying friends have helped make me a better pilot and prompted ideas about what to cover in this book. Rafael Silva (my airplane partner),

Dan Brown, and Peter Alberti have the aviation itch even more than I do. Their enthusiasm is contagious.

While they don't know they helped me, Max Trescott, Juan Brown, and Paul Bertorelli influenced this book in many ways. Max is the host of the *Aviation News Talk* podcast. His content is invaluable for new (and current) pilots. I began listening to him from the day of my first flight lesson and found his insights beyond helpful as I strove to become a skilled and safe pilot. If you're not listening to him, you should be. Juan is the host of the *Blancolirio* channel on YouTube. While his content is varied, most of what he does revolves around exploring lessons learned from aircraft accidents. As I mentioned in this book, every pilot should force themselves to learn about what put other pilots in situations where they bent metal on an airplane, hurt themselves and others, or worse. Juan does an unparalleled job of taking complex topics and explaining them in a clear way that really drives home important lessons. Watch him! Paul was mentioned in a footnote in these pages but is no footnote when it comes to aviation journalism. You can read his work on AVweb.com and in other aviation publications. He also has a large body of work on the AVweb YouTube channel that is at once detailed, informative, and entertaining.

My father, as usual, helped very much. His own writing skill (he's written several books) and his understanding of aviation were critical as I put my own story to paper. And, of course, he deserves all the credit for infecting me with the aviation bug in the first place. I love you, Dad!

During the writing of this book my son, Sean, was taking flight lessons and conversations with him during his training helped me think through what might be included in this book. Thank you, Sean! I Love you son, and I'm proud of you.

Finally, my wife, Mary, was willing to be an "author widow" as I used large chunks of my weekends to write what you have read. Fortunately, as you've by now learned, I got some time to fly with her regularly while I was writing. Her patience and willingness to get into a cockpit with me were so instrumental in allowing me not to just write this book but to pursue my aviation dream. I love you too Mary!

About the Author

Patrick Mullane is an author, professional speaker, and business executive. He graduated from the University of Notre Dame in 1990 with a degree in Mathematics and was commissioned a second lieutenant in the U.S. Air Force. He spent nearly five years in the Air Force, working in an organization that operated intelligence-gathering satellites. Later, he attended Harvard Business School where he received his M.B.A. in 1999. Patrick currently manages an executive education organization and lives in the Boston area with his wife, Mary. Between the two of them, Mary and Patrick have three grown children—Sarah, Sean, and Katie—and two furry friends, Tucker and Finnegan.

Contact me: patrick@pjmullane.com

Printed by Amazon Italia Logistica S.r.l.
Torrazza Piemonte (TO), Italy